Studies in Systems, Decision and Control

Volume 621

The series "Studies in Systems, Decision and Control" (SSDC) covers both new developments and advances, as well as the state of the art, in the various areas of broadly perceived systems, decision making and control–quickly, up to date and with a high quality. The intent is to cover the theory, applications, and perspectives on the state of the art and future developments relevant to systems, decision making, control, complex processes and related areas, as embedded in the fields of engineering, computer science, physics, economics, social and life sciences, as well as the paradigms and methodologies behind them. The series contains monographs, textbooks, lecture notes and edited volumes in systems, decision making and control spanning the areas of Cyber-Physical Systems, Autonomous Systems, Sensor Networks, Control Systems, Energy Systems, Automotive Systems, Biological Systems, Vehicular Networking and Connected Vehicles, Aerospace Systems, Automation, Manufacturing, Smart Grids, Nonlinear Systems, Power Systems, Robotics, Social Systems, Economic Systems and other. Of particular value to both the contributors and the readership are the short publication timeframe and the world-wide distribution and exposure which enable both a wide and rapid dissemination of research output.

Indexed by SCOPUS, DBLP, WTI Frankfurt eG, zbMATH, SCImago.

All books published in the series are submitted for consideration in Web of Science.

Shyam Kamal · Parijat Prasun

A Minimum Function-Based Discrete Sliding Mode Control: Theory and Applications

Springer

Shyam Kamal
Department of Electrical Engineering
Indian Institute of Technology (BHU)
Varanasi
Varanasi, Uttar Pradesh, India

Parijat Prasun
Department of Electrical Engineering
Indian Institute of Technology (BHU)
Varanasi
Varanasi, Uttar Pradesh, India

ISSN 2198-4182 ISSN 2198-4190 (electronic)
Studies in Systems, Decision and Control
ISBN 978-981-95-1160-0 ISBN 978-981-95-1161-7 (eBook)
https://doi.org/10.1007/978-981-95-1161-7

This Springer imprint is published by the registered company Springer Nature Singapore Pte Ltd.
The registered company address is: 152 Beach Road, #21-01/04 Gateway East, Singapore 189721, Singapore

To our family (Aprajita, Adhita, Avyaan, Shushila and Hemant Naithani, Jayanti Naithani), whose patience, understanding, and unwavering support sustained us through the countless hours dedicated to the pursuit of knowledge. Your belief in our abilities strengthened our resolve to overcome challenges and persist in the face of complexity.

Preface

Sliding Mode Control (SMC), first developed in the 1950s for continuous-time systems (CTS), has established itself as a robust technique renowned for its effectiveness in handling nonlinearities, time-varying dynamics, and uncertainties [1–5]. Demonstrating robustness against matched disturbances, SMC offers efficient implementation [6–14]. In control engineering applications where desired performance is required despite disparities between mathematical models and true plant dynamics, in the presence of exogenous perturbations, sliding mode control emerges as a promising technique [4, 6, 15]. Its application has naturally extended to discrete-time systems (DTS) because of the prevalence of digital platforms such as micro-controllers. The core idea of SMC involves selecting a sliding surface to constrain the state's motion, fundamentally shaping the dynamic characteristics of the closed-loop system [4, 16]. A standout attribute of SMC is its inherent robustness against matched bounded disturbances, making it particularly appealing for applications where system dynamics are subject to external perturbations or uncertainties [4, 7].

While sliding mode control is effective, it faces challenges such as chattering [17, 9], characterized by high-frequency oscillations around the sliding surface. Researchers actively seek techniques to mitigate chattering and enhance the practical applicability of sliding mode control [8]. Chattering occurs because of the idealized attribute of the sliding mode, where the system state should precisely follow the sliding surface with instantaneous and abrupt control inputs. However, achieving such idealized motion in practical systems is hindered by physical limitations, actuator dynamics, and the discrete nature of control implementations. Over the years, numerous methods have been developed to alleviate chattering effects [18, 19].

In the realm of practical implementation, many authors have explored discrete-time sliding mode control (DSMC) in the era of digital computers and microcontrollers [20–31], among others. The key distinction between DSMC and continuous-time sliding mode control (CSMC) lies in DSMC's finite switching frequency, eliminating the invariance property inherent in CSMC. DSMC has been explored along two main directions: reaching law-based DSMC, incurring chattering due to the explicit switching term [26, 33], and equivalent control-based DSMC, omitting the reaching phase but potentially requiring substantial control effort [22, 27, 33–36]. The notion of quasi-sliding mode (QSM) had been coined for discrete-time systems because of the impracticality of achieving ideal sliding mode. Additionally, properties of CSMC for DTS differ from those for CTS. Finite-time stability (FTS) for DTS gained attention in recent years [37–39]. The minima-based approach for achieving finite-time stability for DTS was discussed in [37], and later this formulation was extended to convex optimization in [40]. This theory was further explored for discrete sliding mode in [41–43], with Lyapunov characterization in [44] and for twisting-like and super-twisting-like algorithm in discrete-time framework [44, 45].

In the reaching law approach [25–28, 46–52], the next time step's required expression for the switching variable is determined by the reaching law, based on the current switching variable. This methodology is well-suited for discrete-time systems, ensuring a guaranteed sliding mode. Originating from continuous-time systems in [32], this approach extended to DTS in [25]. To address non-ideal sliding motion in DTS, the concept of QSM was introduced and further investigated in [26], incorporating additional conditions for its existence. Gao's reaching law has maintained popularity for almost three decades, serving as a convenient choice for researchers in numerous fields [53]. Even though Gao's reaching law is widely used, it has some specific drawbacks. A notable issue is the coefficient of its proportional term that remains constant, which can lead to a large increments in the switching function, especially under significantly big initial conditions. This can result in an excessively large control action, potentially violating constraints on closed-loop signals. In order to overcome these drawbacks, several modifications to Gao's RL have been proposed, as discussed in studies like [28, 46, 52]. These adaptations concentrate on having the proportional term better, making certain that the design parameters continue to limit the sliding variable's pace of change, thereby offering a more flexible, user-defined solution.

Numerous strategies have been explored to alleviate chattering in discrete-time sliding mode control, as evident in multiple studies. In [27], the focus is on quasi-sliding mode control, aiming to keep states near the switching manifold without recrossing, thereby preventing chattering. The work in [54] introduces a robust discrete sliding mode algorithm to ensure chattering-free operation while effectively managing parameter uncertainties and external disturbances. The study in [55] explores non-smooth DSMC as a strategy to address chattering induced by reaching law-based control. Additionally, the benefits of chattering-free DSMC are illustrated through a significant case study involving missile guidance systems, as showcased in [82].

In the last decade, extensive research has unfolded in the domain of sliding mode control in discrete time, spanning various domains [46, 53, 57–59]. Noteworthy contributions encompass investigations into discrete-time integral SMC tailored for sampled-data systems [62], the exploration of discrete-time terminal SMC [82], and its practical application in addressing perturbations in automotive electronic throttle systems [61]. Advancements in DSMC methodologies include the development of generalized reaching law-based DSMC incorporating variable functions [47], exploring different convergence rates for them [48, 62]. Additionally, studies have delved into band approach analysis, considering predefined band sizes [49], robustness enhancement through relative degree two switching variables [50], its generalization for arbitrary relative degrees [51], and the implementation of reference switching profile-based DSMC [52]. These diverse research endeavors contribute significantly to the evolving landscape of DSMC applications and methodologies.

The twisting algorithm, like the super-twisting algorithm, is extensively utilised in control-related applications because of its straightforward parameter configuration and robustness against model uncertainties. The research community has given it a lot of attention because of its stability, well-posedness, and important qualitative characteristics, like tracking accuracy and convergence rate [19]. In contrast to the super-twisting algorithm, the twisting algorithm exhibits prominent numerical chattering due to its composition of two cascaded signum functions and the absence of an integration operation. Mitigating this numerical chattering has been explored by the sliding mode community, encompassing discrete-time methodologies [55, 63–65] and continuous-time techniques, including the use of higher-order sliding mode (HOSM) [19].

Implicit-Euler methods, as demonstrated in [66–68], have shown significant potential in reducing numerical chattering in contrast to traditional discrete-time implementations that rely on explicit Euler methods. In [69], an implicit-Euler-based technique was proposed to address numerical chattering in the twisting algorithm, with stability and robustness analyzed utilizing the discrete-time Lyapunov method. However, this approach focuses solely on ideal discretization with the implicit-Euler method, neglecting the zero-order hold (ZOH) effect introduced by analog-to-digital converters in practical applications.

Recognizing the influence of the ZOH effect, [70] introduced an implicit-Euler implementation that accounts for ZOH discretization. However, this approach necessitates the use of numerical solvers to resolve the affine variational inequality when computing the control input for the twisting algorithm. This requirement can present challenges for deployment in computation-limited embedded systems and may hinder real-time responsiveness. Furthermore, to address this issue, an implicit Euler zero-order hold discretization without numerical solvers was proposed in [71]. It is crucial to note that all these works focus on discretizing their continuous-time counterpart, intending to realize the continuous-time twisting algorithm using digital controllers.

The super-twisting algorithm (STA), introduced by Levant in 1993 [19], has become a widely embraced algorithm in robust nonlinear control and observation. Operating as a second-order sliding mode algorithm, it is tailored for systems with relative degree one concerning a specified switching function, typically the output. A notable theoretical feature is its capability for finite-time convergence of the output even when matched Lipschitz disturbances are present. In contrast to the sliding mode of first-order controllers, the STA has the advantage of generating a continuous control law, a much needed trait in numerous circumstances. From a practical perspective, the super-twisting algorithm offers straightforward implementation in a digital environment. In existing literature, Euler discretization, both implicit and explicit, is commonly employed to develop discrete-time higher-order sliding mode [82, 63, 64, 68, 69, 71–76]. The explicit Euler discretization method is widely used to implement STA digitally [36, 73, 75, 76], maintaining asymptotic accuracies observed in the continuous-time system.

Among these works, the implicit Euler discretization-based STAs [64, 68, 69] have demonstrated superior chattering suppression compared to explicit Euler discretization methods. The implicit Euler discretization-based super-twisting algorithm (STA) offers unconditional stability, allowing it to address a broader variety of issues without having to worry about numerical instability. In contrast, the explicit Euler discretization is only conditionally stable and can result in numerical instabilities under certain conditions [68]. Furthermore, implicit Euler discretization-based STA has demonstrated greater robustness to noise and uncertainties, higher accuracy with larger sampling times, and improved handling of nonlinearities compared to its explicit counterpart [69].

Unlike explicit Euler discretization, the implicit method is accurate in the unperturbed instance, ensuring the controlled variable converges precisely to zero. Even under disturbances, the state retain the optimal asymptotic accuracy characteristic of the continuous-time system. A key benefit of the implicit method is that its precision remains unaffected by increases in control gains. However, it is important to recognize that discretizing continuous-time sliding mode algorithms could result in periodic behavior [77–79], or, in more severe instances, instability in the closed-loop system [80, 81].

To achieve this objective, developing algorithms directly within the discrete domain, bypassing traditional discretization methods, presents a promising direction for research.

This monograph is driven by the ambition to thoroughly investigate and advance the field of discrete-time sliding mode design based on difference equations with minima for dynamical systems. The motivation behind this exploration lies in the recognition of limitations inherent in classical and conventional control techniques, prompting a quest for methodologies that can effectively address the evolving demands of modern applications. The overarching objective of this monograph is to introduce innovative approaches capable of enabling timely and precise control, even in the presence of uncertainties and exogenous perturbations. Through the development of sophisticated discrete control strategies designed to ensure finite-time convergence, this research aims to enhance the capabilities and operational efficiency of diverse engineering systems, thereby contributing to the ongoing evolution of control methodologies in response to contemporary challenges.

The monograph presents the authors' work on discrete-time sliding mode control design based on the minimum operator. The initial two chapters aim to provide a foundational understanding of discrete sliding mode control and its historical development. Chapter 2 particularly focuses on the necessity of discrete sliding mode control, exploring various design strategies such as switching-based and non-switching-based approaches. The advantages and disadvantages of these strategies are thoroughly examined. The definitions of sliding mode and quasi-sliding mode are elaborated upon. Additionally, a brief discussion is included on alternative approaches for synthesizing reaching laws for discrete-time systems to address the limitations of the aforementioned strategies, with detailed exploration in subsequent chapters. The subsequent sections of the monograph encompass contributory chapters, summarized as follows:

- The discussion begins with a detailed examination of the history and evolution of discrete-time sliding mode control, rooted in the fundamental principles of continuous-time sliding mode control. A basic design methodology for CSMC is first introduced, followed by an analysis of the development of DSMC, with emphasis on the key differences in their design approaches. Furthermore, a significant classification is presented, distinguishing between switching and non-switching reaching law developments within these control frameworks. These foundational concepts are essential for establishing the context and rationale behind the research contributions presented in the subsequent chapters.
- Chapter 4 marks our initial exploration where we introduced two new reaching laws designed specifically for discrete-time systems, utilizing a difference equation with minima. These reaching laws not only effectively eliminate chattering, ensuring a smoother operation, but also contribute to improved system robustness. This enhancement is achieved by strategically reducing the width of the quasi-sliding mode band without placing excessive stress on the control effort. The delicate balance achieved by these reaching laws highlights their effectiveness in addressing challenges within discrete-time systems. The proposed methodology's effectiveness and robustness is validated through theoretical analysis and extensive simulation studies, showcasing its capability to achieve sliding mode for the unperturbed case and quasi-sliding mode for the perturbed case.
- Following the discussion on reaching laws based on the difference equation with minima, in Chap. 5, our exploration extended to incorporating rate-regulatory function-based reaching laws. The primary aim is to harness the advantages associated with the variable pace at which the switching function changes, achieved through the incorporation of a variable gain function in the proportional term. This modification leads to a significantly accelerated rate of change of the switching function. Notably, we demonstrate that this rate of change is effectively upper-bounded by the design parameters, introducing an additional advantageous feature. The proposed theory is validated via simulations that include a numerical example and a pendulum system. In the case of the pendulum system, a comparative analysis among various methods is provided which consistently highlight the superior performance of the proposed approach.
- Moreover, Chap. 6 extensively explores the Lyapunov characterization of discrete-time sliding mode control based on a difference equation with minima. This involve formulating Lyapunov conditions for the finite-time input-to-state stability of perturbed discrete-time systems along with providing insights into the settling time function. Furthermore, we delve into the Lyapunov stability analysis of the closed-loop system. Additionally, we conduct a comparative study through simulations, highlighting the superior performance of the proposed laws in comparison to existing ones. An intriguing observation arising from this analysis is the achievement of the least ultimate bound for the sliding variable trajectory along with satisfactory transient behavior.

- In Chap. 7, we shift our focus towards the development of discrete-time twisting-like and super-twisting-like algorithms, grounded in the framework of a difference equation with minima. Diverging from the conventional Euler-discretization often employed in discrete-time algorithms, our proposed designs showcase the notable distinctiveness. Through the application of these innovative designs, we successfully achieved the convergence of system states to the origin in the case of unperturbed systems and to an invariant set for systems with disturbances, all within finite-time span. Additionally, we extended the utility of the discrete-time super-twisting-like algorithm to design a super-twisting observer. This observer, in a significant development, robustly estimate unknown state variables of the system within a finite timeframe, highlighting its effectiveness.
- In the final chapter, our focus shifted towards the practical application of the proposed minimum function based reaching law with difference equation framework in benchmark setups, providing a detailed exploration of the applied reaching law. This real-world implementation offer a comprehensive analysis of the reaching law's performance within the intricate dynamics of the practical systems. The obtained results not only showcased the robustness of the proposed approach but also highlighted its adaptability to real-world experimental conditions. The successful implementation of the proposed law in the these systems is meticulously demonstrated, presenting the outcomes through a combination of simulation and experimentation.

Varanasi, India
November 2024

Shyam Kamal
Parijat Prasun

References

1. Emelyanov, S.V.: "A method to obtain complex regulation laws using only the error signal or the regulated coordinate and its first derivatives," Avtomat. Telemekh., **18**(10), 873–885 (1957)
2. Utkin, V.I.: "Variable structure systems with sliding modes," IEEE Trans. Autom. Control, **22**(2) 212–222, (1977)
3. Utkin, V.I.: "Sliding Modes in Control and Optimization," Berlin, Germany: Springer-Verlag, (1992)
4. Edwards, C., Spurgeon, S.: "Sliding mode control: theory and applications," CRC Press, (1998)
5. Posielek, T., Wulff, K., Reger, J.: "Analysis of sliding-mode control systems with relative degree altering perturbations," Automatica, 148, 110745, (2023)
6. Young, K.D., Utkin, V.I., Ozguner, U.: "A control engineer's guide to sliding mode control," IEEE transactions on control systems technology, **7**(3), 328–342 (1999)
7. Perruquetti, W., Barbot, J.P.: "Sliding mode control in engineering," CRC press, (2002)
8. Laghrouche, S., Plestan, F., Glumineau, A.: "Higher order sliding mode control based on integral sliding mode," Automatica **43**(3) 531–537, (2007)
9. Shtessel, Y., Edwards, C., Fridman, L., Levant, A.: "Sliding Mode Control and Observation," 1st ed., New York: Springer, Birkhauser, (2014)
10. Kumari, K., Bandyopadhyay, B., Reger, J., Behera, A.K.: "Event-triggered sliding mode control for a high-order system via reduced-order model based design," Automatica, **121**, 109163, (2020)

11. Thomas, M., Bandyopadhyay, B.: "Discrete-time sliding mode control design for unicycle robot with bounded inputs", IEEE Trans. Circuits and Systems II: Express Briefs, **68**(8), 2912–2916 (2021)
12. Pal, A.K., Kamal, S., Yu, X., Nagar, S.K., Bandyopadhyay, B.: "Free-will arbitrary time terminal sliding mode control", IEEE Trans. Circuits and Systems II: Express Briefs, (2020)
13. Singh, B., Pal, A.K., Kamal, S., Dinh, T.N., Mazenc, F.: "Vector control Lyapunov function based stabilization of nonlinear systems in predefined time," IEEE Trans. on Autom. Control, (2022)
14. Zhang, L., Chen, Z., Yu, X., Yang, J., Li, S.: "Sliding-mode-based robust output regulation and its application in PMSM servo systems," IEEE Transactions on Industrial Electronics, **70**(2), 1852–1860 (2023)
15. Utkin, V.I., Vadim, I.: "Sliding mode control," Variable structure systems: from principles to implementation **66** (2004)
16. Fridman, L., Moreno, J.A., Bandyopadhyay, B., Kamal, S., Chalanga, A.: "Continuous nested algorithms: The fifth generation of sliding mode controllers," in Recent advances in sliding modes: From control to intelligent mechatronics, 5–35, Springer, (2015)
17. Lee, H., Utkin, V.I.: "Chattering suppression methods in sliding mode control systems," Annual reviews in control **31**(2), 179–188 (2007)
18. Plestan, F., Glumineau, A., Laghrouche, S.: "A new algorithm for high-order sliding mode control," International Journal of Robust and Nonlinear Control, **18**(4–5), 441–453, (2008)
19. Levant, A.: "Sliding order and sliding accuracy in sliding mode control," Int. Jour. Control, **58**(6), 1247–1263 (1993)
20. Milosavljevic, D.: "General conditions for the existence of a quasi-sliding mode on the switching hyperplane in discrete variable structure systems," Autom. Remote Control, **46**(3), 307–314, (1985)
21. Sarpturk, S., Istefanopulos, S.Y., Kaynak, O.: "On the stability of discrete-time sliding mode control systems," IEEE Trans. Automat. Contr., **32**, 930–932 (1987)
22. Utkin, V.I., Drakunov, S.V.: "On discrete-time sliding modes," IFAC Proc. Volumes, **22**(3), 273–278 (1989)
23. Furuta, K.: "Sliding mode control of a discrete system," Systems and Control Letters, **14**(2), 145–152 (1990)
24. Sira-Ramirez, H.: "Non-linear discrete variable structure systems in quasi-sliding mode," Inter. Jour. control, **54**(5), 1171–1187 (1991)
25. Gao, W., Wang, Y., Homaifa, A.: "Discrete-time variable structure control systems," IEEE Trans. on Ind. Electron., **42**(2), 117–122 (1995)
26. Bartoszewicz, A.: "Remarks on discrete-time variable structure control systems," IEEE Trans. on Ind. Electron., **43**(1), 235–238 (1996)
27. Bartoszewicz, A.: "Discrete-time quasi-sliding-mode control strategies," IEEE Trans. on Ind. Electron., **45**(4), 633–637 (1998)
28. Bartoszewicz, A., Lesniewski, P.: "New switching and nonswitching type reaching laws for SMC of discrete time systems," IEEE Trans. Control Syst. Tech., **24**(2), 670–677 (2016)
29. Zhang, J., Shi, P., Xia, Y., Yang, H.: "Discrete-time sliding mode control with disturbance rejection," IEEE Trans. Indus. Electr., **66**(10), 7967–7975 (2018)
30. Hettiger, C., Watermann, L., Kumari, K., Eisenzopf, L., Weissenberger, F., Horn, M., Koch, S., Reger, J., Reichhartinger, M.: "On Discretization Methods for Indirect Adaptive Sliding Mode Control," IEEE 61st Conference on Decision and Control (CDC), 4930–4936, (2022)
31. Michel, N., Olaru, S., Valmorbida, G., Bertrand, S., Dumur, D.: "Invariant sets for discrete-time constrained linear systems using a sliding mode approach," 2018 European Control Conference (ECC), 2929–2934, (2018)
32. Gao, W., Hung, J.C., "Variable structure control of nonlinear systems: A new approach," IEEE Trans. on Ind. Electron., **40**(1), 45–55 (1993)
33. Utkin, V.I.: "Sliding mode control in discrete-time and difference systems," Variable structure and Lyapunov control, 87–107 (1994)

34. Su, W.C., Drakunov, S.V., Ozguner, U.: "An $O(T^2)$ boundary layer in sliding mode for sampled-data systems," IEEE Trans. on Autom. Control, **45**(3), 482–485 (2000)
35. Yu, X., Chen, G.: "Discretization behaviors of equivalent control based sliding-mode control systems," IEEE Trans. on Autom. Control **48**(9), 1641–1646 (2003)
36. Galias, Z., Yu, X.: "Euler's discretization of single input sliding-mode control systems," IEEE Trans. on Autom. Control **52**(9), 1726–1730 (2007)
37. Haddad, W.M., Lee, J.: "Finite-time stability of discrete autonomous systems," Automatica, **122**, (2020)
38. Lee, J., Haddad, W.M., Bhat, S.P.: "Finite time stability of discrete-time stochastic dynamical systems," 60th IEEE Conf. Deci. Control, 6646–6651. IEEE, (2021)
39. Haddad, W.M., Lee, J.: "Finite-time stabilization and optimal feedback control for nonlinear discrete-time systems," IEEE Trans. Autom. Control, https://doi.org/10.1109/TAC.2022.3151195, (2022)
40. Prasun, P., Pandey, S., Kamal, S., Ghosh, S., Singh, D.: "Discrete-time gradient systems governed by difference equation with minima," 31st Mediterranean Conf. Contr. and Autom., Limassol, Cyprus, (2023)
41. Prasun, P., Kamal, S., Bartoszewicz, A., Ghosh, S.: "A difference equation with minima-based reaching law for discrete variable structure systems," IEEE Trans. Circ. and Syst. II: Expr. Briefs, **71**(3), 1236–1240 (2023)
42. Prasun, P., Kamal, S., Bartoszewicz, A., Ghosh, S.: "Difference equation with minima based discrete-time sliding mode control," IEEE Trans. Circ. and Syst. II: Expr. Briefs, **70**(12), 4404–4408 (2023)
43. Prasun, P., Kamal, S., Ghosh, S., Dinh, T.N.: "A minimum operator based discrete variable structure control," European Journal of Control, **76**, 100953 (2024)
44. Prasun, P., Kamal, S., Pandey, S., Bartoszewicz, A., Ghosh, S.: "A minimum operator based discrete-time sliding mode control," IEEE Transactions on Automatic Control, **69**(11), 7871–7876 (2024)
45. Prasun, P., Pandey, S., Kamal, S., Ghosh, S., Xiong, S.: "A minimum operator based discrete-time super-twisting-like algorithm," IEEE Trans. Circ. and Syst. II: Expr. Briefs, **71**(1), 286–290, (2023)
46. Bartoszewicz, A., Latosiński, P.: "Discrete time sliding mode control with reduced switching–a new reaching law approach," International Journal of Robust and Nonlinear Control, **26**(1), 47–68 (2016)
47. Chakrabarty, S., Bandyopadhyay, B.: "A generalized reaching law for discrete time sliding mode control," Automatica, **52**, 83–86 (2015)
48. Chakrabarty, S., Bandyopadhyay, B.: "A generalized reaching law with different convergence rates," Automatica, **63**, 34–37 (2016)
49. Chakrabarty, S., Bandyopadhyay, B.: "Minimum ultimate band design of discrete sliding mode control," Asian journal of control, **17**(5), 1889–1897 (2015)
50. Chakrabarty, S., Bartoszewicz, S.: "Improved robustness and performance of discrete time sliding mode control systems," ISA transactions, **65**, 143–149 (2016)
51. Bartoszewicz, A., Latosiński, P.: "Generalization of Gao's reaching law for higher relative degree sliding variables," IEEE Trans. on Autom. Control, **63**(9), 3173–3179 (2018)
52. Bartoszewicz, A., Adamiak, K.: "Discrete-time sliding-mode control with a desired switching variable generator," IEEE Transactions on Automatic Control, **65**(4), 1807–1814 (2020)
53. Kurode, S., Bandyopadhyaya, B., Gandhi, P.S.: "Discrete sliding mode control for a class of underactuated systems," 37th Annu. Conf. IEEE Ind. Electron. Soc. 3936–3941, (2011)
54. Golo, G., Milosavljević, Č.: "Robust discrete-time chattering free sliding mode control," Systems and Control Letters, **41**(1), 19–28 (2000)
55. Du, H., Yu, X., Chen, M.Z.Q., Li, S.: "Chattering-free discrete-time sliding mode control," Automatica, **68**, 87–91, (2016)
56. Wang, X., Wu, D., He, J., Zhang, J., Li, S.: "Chattering-free discrete-time sliding mode control for integrated missile guidance and control system," Proceedings of the Institution of Mechanical Engineers, Part G: Journal of Aerospace Engineering, **237**(7), 1546–1556 (2023)

57. Sharma, N.K., Roy, S., Janardhanan, S., Kar, I.N.: "Adaptive discrete-time higher order sliding mode," IEEE Trans. Circuits and Systems II: Express Briefs, **66**(4), 612–616 (2018)
58. Latosinski, P., Bartoszewicz, A.: "Model reference DSMC with a relative degree two switching variable," IEEE Trans. Autom. Control, **66**(4), 1749–1755 (2020)
59. Mishra, J.P., Yu, X.: "Delta-operator-based reaching laws for sliding mode control design," IEEE Trans. Circuits and Systems II: Express Briefs, **69**(4), 2136–2140 (2021)
60. Abidi, K., Xu, J.X., Yu, X.: "On the discrete-time integral sliding-mode control," IEEE Trans. on Autom. Control, **52**(4), 709–715 (2007)
61. Song, M., Li, S., Du, H., Wang, Z.: "Chattering-free discrete-time fast terminal sliding mode control of automotive electronic throttle with disturbances," 39th Chinese Control Conference, Shenyang, China, (2020)
62. Ren, Q., Kao, Y., Zhu, Y., Xia, H., Wang, C.: "A new discrete reaching condition and generalized discrete reaching law with different convergence rates," Automatica, **132**, (2021)
63. Acary, V., Brogliato, B., Orlov, Y.V.: "Chattering-free digital sliding-mode control with state observer and disturbance rejection," IEEE Trans. Autom. Control, **57**(5), 1087–1101, (2011)
64. Xiong, X., Pal, A.K., Liu, Z., Kamal, S., Huang, R., Lou, Y.: "Discrete-time adaptive super-twisting observer with predefined arbitrary convergence time," IEEE Trans. Circ. and Syst. II: Expr. Briefs, **68**(6), 2057–2061 (2021)
65. Yan, Y., Galias, Z., Yu, X., Sun, C.: "Euler's discretization effect on a twisting algorithm based sliding mode control," Automatica, **68**, 203–208 (2016)
66. Huber, O., Acary, V., Brogliato, B.: "Lyapunov stability analysis of the implicit discrete-time twisting control algorithm," IEEE Trans. on Autom. Control, **65**(6), 2619–2626 (2020)
67. Huber, O., Acary, V., Brogliato, B.: "Lyapunov stability analysis of the implicit discrete-time twisting control algorithm," IEEE Trans. on Autom. Control, **65**(6), 2619–2626 (2020)
68. Brogliato, B., Polyakov, A., Efimov, D.: "The Implicit discretization of the supertwisting sliding-mode control algorithm," IEEE Trans. Autom. Control, **65**, 3707–3713 (2020)
69. Xiong, X., Kikuuwe, R., Kamal, S., Jin, S.: "Implicit-Euler implementation of super-twisting observer and twisting controller for second-order systems," IEEE Trans. on Circuits and Systems II: Express Briefs, **67**(11), 2607–2611 (2020)
70. Huber, O., Acary, V., Brogliato, B., Plestan, F.: "Implicit discrete-time twisting controller without numerical chattering: Analysis and experimental results," Control Eng. Practice, **46**, 129–141 (2016)
71. Xiong, X., Bai, Y., Shi, R., Kamal, S., Wang, Y., Lou, Y.: "Discrete-Time Twisting Algorithm Implementation with Implicit-Euler ZOH Discretization Method," IEEE Trans. on Circuits and Systems II: Express Briefs, **69**(8), 3435–3439 (2022)
72. Salgado, I., Kamal, S., Chairez, I., Bandyopadhyay, B., Fridman, L.: "Super-twisting-like algorithm in discrete time nonlinear systems," in The 2011 Inter. Conf. on Advanced Mechatronic Systems, 497–502 (2011)
73. Salgado, I., Kamal, S., Bandyopadhyay, B., Chairez, I., Fridman, L.: "Control of discrete time systems based on recurrent super-twisting-like algorithm," ISA Trans., **64**, 47–55 (2016)
74. Koch, S., Reichhartinger, M.: "Discrete-time equivalents of the super-twisting algorithm," Automatica, **107**, 190–199 (2019)
75. Salgado, I., Chairez, I., Bandyopadhyay, B., Fridman, L., Camacho, O.: "Discrete-time nonlinear state observer based on a super twisting-like algorithm," IET Contr. Theo. & Appli., **8**, 803–812 (2014)
76. Yan, Y., Yu, X., Sun, C.: "Discretization behaviors of a super-twisting algorithm based sliding mode control system," Int. Works. Recent Adva. Sliding Modes (RASM), 1–5 (2015)
77. Galias, Z., Yu, X.: "Study of periodic solutions in discretized two-dimensional sliding-mode control systems," IEEE Trans. Circuits and Systems II: Express Briefs, **58**(6), 381–385 (2011)
78. Xia, X., Zinober, A.S.I.: "Δ-Modulated feedback in discretization of sliding mode control," Automatica, **42**, 771–776 (2006)
79. Levant, A., Livne, M., Lunz, D.: "On discretization of high-order sliding modes", in Recent Trends Sliding Mode Cont., **102**, 177 (2016)

80. Janardhanan, S., Bandyopadhyay, B.: "On discretization of continuous-time terminal sliding mode," IEEE Trans. Autom. Control, **51**, 1532–1536 (2006)
81. Behera, A.K., Bandyopadhyay, B.: "Steady-state behaviour of discretized terminal sliding mode," Automatica, **54**, 176–181 (2015)
82. Li, S., Du, H., Yu, X.: "Discrete-time terminal sliding mode control systems based on Euler's discretization," IEEE Trans. on Autom. Control **59**(2), 546–552 (2013)

Acknowledgements

While our name stands alone on the cover of this monograph, we are keenly aware that its creation wouldn't have been possible without the support and contributions of numerous exceptional individuals. Our sincere gratitude extends to all those who have played a substantial role in transforming this monograph into a reality. Their guidance, encouragement, and assistance have molded our experience into an unforgettable journey that we will forever hold dear. We take this opportunity to express our deep gratitude and utmost respect to individuals from the Department of Electrical Engineering at IIT (BHU) Varanasi. Their consistent oversight, and unwavering encouragement have played a pivotal role throughout the entire journey of this monograph. We are genuinely thankful for their support, which has been a significant contributing factor to the successful completion of this work. In particular, we extend profound gratitude to Dr. Sandip Ghosh for his invaluable guidance and steadfast support. We also acknowledge the support received from our colleagues and all the members of Control Systems Research Lab.

We want to express our sincere appreciation to Prof. Andrzej Bartoszewicz from Lodz University of Technology, Poland, for offering crucial insights and invaluable suggestions during the drafting of certain manuscripts associated with my monograph. Additionally, we want to thank Dr. Debdas Ghosh, Department of Mathematical Sciences, IIT (BHU) Varanasi, for his valuable mathematical suggestions which has improved our work. Gratitude is also extended to Dr. Thach N. Dinh, Associate Professor from Conservatoire National des Arts et Metiers (CNAM), France, and Dr. Xiaogang Xiong, Associate Professor from HIT Shenzhen, for their collaborative efforts, insightful suggestions, and valuable advice, all of which have significantly elevated the quality of our research. Your support and input have played a pivotal role in shaping the outcomes of our work, and we are truly grateful for your substantial contributions.

Finally, and certainly of immense significance, we wish to convey our profound appreciation to our parents, and family members, for their steadfast support and encouragement throughout this endeavor. Their love and counsel have served as our beacon, and we are eternally grateful to them. Their unwavering belief in our capabilities has been a wellspring of resilience, particularly during moments of adversity.

Contents

About the Authors

Shyam Kamal received his Bachelor's degree in Electronics and Communication Engineering from the Gurukula Kangri Vishwavidyalaya Haridwar, Uttarakhand, India in 2009, and Ph.D. in Systems and Control Engineering from the Indian Institute of Technology Bombay, India in 2014. From 2015 to 2016, he was with the Department of Systems Design and Informatics, Kyushu Institute of Technology, Japan as a Project Assistant Professor. Currently, he is an Associate Professor in the Department of Electrical Engineering, Indian Institute of Technology (BHU) Varanasi, India. He has published one monograph, two book chapters and 105 journal articles and conference papers. His research interests include the areas of nonlinear control and its applications. Particularly he is working on fractional order systems, contraction analysis, discrete and continuous higher-order sliding mode control and multiagent systems. He has received excellence awards in Ph.D. thesis from IIT Bombay in 2015 for his thesis "Sliding Mode Control of Fractional-order Systems" and INAE Young Engineer Award in 2019 for his research work. He is now an INAE Young Associate. He has also received honorary appointment as a Visiting professor from RMIT Melbourne Australia in July 2017 and in the duration of November and December 2019 and Visiting Professor from Harbin Institute on Technology China in August 2019.

Parijat Prasun obtained his B.Tech. degree in Electrical and Electronics Engineering from the College of Engineering Roorkee, Uttarakhand, India, in 2015. He earned his M.Tech. degree in Electrical Engineering with a specialization in Power and Energy Systems from the National Institute of Technology Meghalaya, India, in 2018. He later completed his Ph.D. in Control Systems at the Indian Institute of Technology (BHU) Varanasi, India, in 2024. His research interests include nonlinear control, discrete sliding mode control, and finite- and fixed-time control design.

Acronyms

Abbreviations

CSMC	Continuous-Time Sliding Mode Control
CTS	Continuous-Time System
DEM	Difference Equation with Minima
DSMC	Discrete-Time Sliding Mode Control/Discrete Sliding Mode Control
DTS	Discrete-Time System
FTISS	Finite-Time Input-to-State Stable
HOSM	Higher-Order Sliding Mode
ISS	Input-to-State Stable
lim	Limit
max	Maximum
min	Minimum
MTBP	Matched Type Bounded Perturbations
PID	Proportional Integral and Derivative
QSM	Quasi-Sliding Mode
RL	Reaching Law
s.t.	Such that
SMC	Sliding Mode Control
STA	Super-Twisting Algorithm
STO	Super-Twisting Observer
VSC	Variable Structure Control
ZOH	Zero-Order Hold

Greek and Roman Symbols

$\mathbb{R}$	Real number set
$\mathbb{R}_+$	Set of positive real numbers

$\mathbb{R}^n$	A vector space of dimension-n whose entries belong to real number set $\mathbb{R}$
$\mathbb{R}_{\geq 0}$	Set of non-negative real numbers
$\mathbb{Z}_{\geq 0}$	Set of non-negative integers
$\mathbb{R}_{\geq k}$	Set of real numbers greater than or equal to k
$\mathbb{R} \to \mathbb{R}$	Element-wise mapping
$\mathcal{K}$	Class $\mathcal{K}$ function
$\mathcal{K}_\infty$	Class $\mathcal{K}_\infty$ function
$\mathcal{L}$	Class $\mathcal{L}$ function
$\mathcal{KL}$	Class $\mathcal{KL}$ function
$\mathcal{GK}$	Generalized-$\mathcal{K}$ function
$\mathcal{GKL}$	Generalized-$\mathcal{KL}$ function

Chapter 1
A Journey from Classical Control Towards Variable Structure Control

Abstract This introductory chapter covers the journey of control system theory research from classical control towards variable structure control. Control system theory is crucial for regulating dynamic systems across various applications, ensuring stability, performance, and robustness against disturbances. It covers both continuous-time and discrete-time systems, each serving different domains. Feedback mechanisms are central to this theory, enabling systems to maintain desired outputs by minimizing discrepancies between actual and reference signals. Conventional control methods, such as PID controllers, rely on continuous actions and frequency domain analysis, excelling in linear time-invariant systems. In contrast, Variable Structure Control (VSC) utilizes discontinuous actions, with Sliding Mode Control (SMC) being a prominent method that effectively handles nonlinearities and uncertainties, despite challenges like chattering. With the rise of digital technology, Discrete-Time Sliding Mode Control (DSMC) has emerged, adapting SMC principles for discrete systems. While DSMC is efficient for digital control, it faces challenges such as quasi-sliding modes, where the system's behavior only approximates ideal sliding motion. To address this, higher-order discrete sliding modes have been developed, offering enhanced robustness and precision by reducing the quasi-sliding mode band. Overall, control system theory provides a versatile framework for analyzing, designing, and optimizing dynamic systems, with ongoing advancements like DSMC and higher-order methods continuing to refine its applicability across various fields.

1.1 Introduction

1.1.1 Basic Notion of Control System

Control system theory is a foundational discipline that plays a pivotal role in regulating and managing the behavior of dynamic systems across a diverse range of applications. It encompasses a comprehensive set of principles, methodologies, and techniques designed to ensure the stability, performance, and robustness of systems

S. Kamal and P. Prasun, *A Minimum Function-Based Discrete Sliding Mode Control: Theory and Applications*, Studies in Systems, Decision and Control 621,
https://doi.org/10.1007/978-981-95-1161-7_1

subject to external influences, disturbances, or uncertainties. At its core, control system theory is concerned with understanding, analyzing, and designing systems that can automatically maintain desired outputs or trajectories. Whether applied in engineering, physics, economics, or biological systems, control theory provides a systematic framework to model, analyze, and optimize the behavior of dynamic systems.

Key concepts within control system theory include feedback, where the system output is compared to a reference signal, and adjustments are made to minimize the discrepency or error. This feedback loop allows for continuous monitoring and correction, ensuring that the system responds appropriately to changing conditions. The study of control systems is broadly categorized into two main types: continuous-time and discrete-time systems. Continuous-time systems deal with signals that vary smoothly over time, while discrete-time systems operate with signals that are sampled at discrete intervals. Both types find applications in various domains, from electrical and mechanical engineering to aerospace, chemical processes, and beyond.

Control system theory involves the use of mathematical models, often in the form of differential equations or difference equations, to represent the dynamics of the system. Stability analysis is a critical aspect, assessing the system's tendency to return to a desired state after perturbations. Additionally, control design methodologies aim to determine the optimal parameters for controllers that govern the system's behavior. The development and application of various control strategies, such as proportional-integral-derivative (PID) controllers, state-space control, and optimal control, further contribute to the versatility and efficiency of control systems.

Further, control theory can be classified into conventional control, predominantly characterized by its continuous nature, and non-conventional control, exemplified by variable structure control, which incorporates some form of structural discontinuity. Conventional control and variable structure control represent distinct methodologies within the realm of control systems, each characterized by its unique principles and applications. Conventional control, rooted in frequency domain analysis, employs tools like Bode plots and Nyquist diagrams to assess stability, and often relies on proportional-integral-derivative controllers for system regulation. It is particularly well-suited for linear time-invariant systems and finds applications in areas such as process control, automotive engineering, etc. In contrast, variable structure control introduces a paradigm shift by utilizing discontinuous control actions and switching between different control structures. Sliding mode control, a common technique in VSC, constrains the system trajectory to a sliding surface, making it adept at handling nonlinearities and uncertainties. Stability analysis in VSC often involves Lyapunov stability theory, emphasizing robustness in the face of uncertainties. The choice between conventional control and variable structure control depends on the nature of the controlled system, with conventional control excelling in linear systems and variable structure control offering robust solutions for nonlinear and uncertain systems, such as those encountered in robotics and complex systems.

1.1.2 Sliding Mode Control: A Paradigm Shift

Sliding mode control is a robust and effective control strategy widely employed in the field of control systems. At its core, SMC operates by creating a sliding surface, a predefined subspace within the system's state space, on which the system's trajectory is forced to converge, thereby inducing a sliding motion. This sliding motion ensures the system continues to be surface-based, minimizing the impact of uncertainties, disturbances, and variations. One of the key strengths of sliding mode control is its inherent ability to handle nonlinearities and external disturbances, making it particularly suitable for systems where uncertainties are prevalent. The control law in sliding mode control is typically discontinuous, introducing a switching element that facilitates robustness and stability. SMC has found applications in diverse fields, including aerospace, robotics, power systems, and automotive control, owing to its ability to provide reliable performance in the face of dynamic uncertainties and varying operating conditions. Despite its effectiveness, the chattering phenomenon, a high-frequency oscillations around the sliding surface, can be a challenge in practical implementations, prompting ongoing research to refine and enhance the methodology for improved real-world applicability.

The widespread accessibility of computing devices and micro-controllers has facilitated the advancement of discrete-time sliding mode control (DSMC). DSMC extends the principles of the well-established sliding mode control framework to discrete-time systems. In this approach, the control law is designed to ensure that the system trajectory adheres to a predefined sliding surface within each sampling interval. This approach is particularly valuable in applications where the system dynamics are inherently discrete or when dealing with digital control systems. DSMC does not inherit the complete disturbance rejection characteristics of its continuous-time counterpart, making it open for exploration and enhance the degree of robustness. The discrete nature of DSMC aligns with the practical constraints of digital control implementations, offering advantages in terms of computational efficiency and ease of implementation in discrete-time systems. We could commonly apply DSMC in fields such as digital control systems, embedded systems, and networked control systems where discrete-time dynamics play a crucial role. The design and analysis of discrete-time sliding mode control involve considerations of the system's discrete-time model, discretization techniques, and the impact of the sampling interval on the overall performance and stability of the control system.

A crucial challenge arising from discretization is the loss of the ideal sliding mode behavior. It follows from this that the sliding variable no longer precisely adheres to the sliding manifold, giving rise to what is known as quasi-sliding mode (QSM). The corresponding region around the sliding manifold, known as the quasi-sliding mode band (QSMB), encapsulates the proximity within which the sliding variable operates.

One way to reduce the size of aforementioned quasi-sliding mode band width is by the use of higher-order discrete sliding modes. Higher-order discrete sliding mode control represents an advanced approach within the realm of discrete-time control systems, specifically tailored for systems influenced by bounded perturbations. This methodology involves increasing the relative degree of the sliding variable, enabling enhanced control over the system's behavior. By doing so, higher-order discrete sliding mode control aims to reduce the size of the quasi-sliding mode band, consequently improving robustness and performance. The twisitng algorithm and super-twisting algorithm are very popular second-order sliding mode in continuous-time counterpart. Over the years, many researchers have explored the discretized versions of aforementioned algorithm's continuous-time counterpart and reported the benefits and properties it retained. This pursuit not only enhances the system's robustness but also contributes to a more precise and efficient control scheme, making higher-order discrete sliding mode control a significant area of interest in advancing control system theories and applications.

1.2 Motivation

Control systems are instrumental in a multitude of engineering applications, spanning domains such as industrial automation, aerospace, transportation, robotics, and energy sectors. Their overarching goal is to influence the behavior of dynamic systems to attain specific performance objectives. Classical control methods like PID controllers and model-based control have undergone thorough examination and found widespread use in diverse fields. Nevertheless, with the progression of technology and the increasing complexity of systems, traditional control approaches are encountering constraints in fulfilling the requirements of contemporary applications. The evolving landscape of engineering challenges necessitates innovative control strategies that can effectively address the intricacies and demands posed by advanced systems and emerging technologies.

Addressing the demand for swift and precise responses within a finite-time framework represents a significant challenge in the design of control systems. Numerous real-world systems necessitate control strategies that not only ensure stability and accuracy but also achieve these objectives within a finite time duration. Classical control methods, such as PID control, typically achieve tracking or stabilization asymptotically, implying that the system never precisely reaches the desired trajectory. In industrial manufacturing, for instance, where precision is crucial, control systems must adhere to stringent timing constraints to optimize production efficiency and uphold quality standards. These instances underscore the imperative for control methodologies capable of ensuring finite-time convergence while meeting specific performance criteria. The quest for control strategies that combine speed, accuracy, and finite-time guarantees reflects the evolving landscape of control system design in response to the demands of diverse and dynamic applications.

This monograph is motivated by the aspiration to delve into and advance the domain of discrete-time sliding mode design based on difference equations with minima for dynamical systems. The impetus behind this exploration stems from the recognition of limitations inherent in classical and conventional control techniques, prompting a quest for methodologies that can effectively meet the evolving demands of modern applications. The overarching goal of this monograph is to introduce innovative approaches capable of facilitating time based and precise control, even in the face of uncertainties and exogenous perturbations. Through the development of sophisticated discrete control strategies designed to ensure finite-time convergence, this research endeavors to elevate the capabilities and operational efficiency of diverse engineering systems, thereby contributing to the ongoing evolution of control methodologies in the face of contemporary challenges.

To be more specific, the exploration of discrete-time sliding mode is primarily driven by the pervasive uncertainties present in real-world systems. These uncertainties may stem from inherent factors within the system, such as parameter variations and modeling errors, or they may manifest externally in the form of disturbances. The influence of uncertainties poses a significant challenge to control performance in practical systems. Traditional control methods often struggle, especially when confronted with time-varying non-vanishing perturbations, leading to suboptimal performance or even instability. In response to these challenges, discrete-time sliding mode control methodologies emerge as a promising avenue. By ensuring convergence within a finite time frame, irrespective of the effects of uncertainties and perturbations, these methodologies present a robust solution for enhancing control performance in the face of real-world complexities.

The majority of existing methodologies in discrete-time sliding mode control are grounded in the reaching law approach, incorporating both switching as well as non-switching terms. Notably, reaching laws based on switching functions often exhibit a significant reaching phase, particularly noticeable for large initial conditions, accompanied by chattering and the presence of a quasi-sliding mode band. The widely adopted Gao's reaching law, utilizing a switching term approach, explicitly introduces chattering, leading to trajectories residing in an ultimate band during steady-state. This approach also suffers from a drawback of a substantial reaching phase for exceptionally large initial conditions. Conversely, the equivalent control-based reaching law, initially proposed by Utkin, takes a non-switching approach. While it results in a dead-beat type response with no reaching phase, it demands a considerable control action, presenting a challenge in practical applications. These fundamental issues persist unresolved, indicating a pressing need for comprehensive solutions to address the challenges associated with discrete-time sliding mode control methodologies.

Furthermore, the sliding mode community has dedicated extensive efforts to the development of discrete-time twisting and super-twisting algorithms tailored for discrete-time dynamical systems. The super-twisting algorithm (STA) stands out as a celebrated second-order sliding mode algorithm. While the theoretical appeal of the continuous-time super-twisting algorithm is evident, its practical implementation faces challenges, particularly in directly extending the concept to discrete-time systems. This challenge arises due to the non-continuous nature of the sliding variable and its higher-order differences in DTS. Various existing designs employ Euler discretization in both explicit and implicit forms, with implicit Euler-discretized versions demonstrating superior chattering suppression compared to explicit Euler-discretized methods.Implicit Euler-discretized super-twisting algorithms (STAs) offer unconditional stability, allowing them to address greater variety of issues without the possibility of numerical instability. In contrast, explicit Euler-discretized methods are stable under certain conditions and susceptible to numerical instabilities. Moreover, studies show that implicit Euler-discretized STAs demonstrate greater robustness to noise and uncertainties, higher accuracy with larger sampling intervals, and better performance in handling nonlinearities. It is crucial to note that the discretization of continuous-time sliding mode algorithms may lead to periodic behavior or, in more severe cases, an unstable closed-loop system. Consequently, the justified necessity to design these algorithms separately in the discrete domain warrants further attention.

As far as the implementation of the discrete-time sliding mode algorithm is concerned, it is greatly influenced by the rapid advancement of computing technologies and the availability of increasingly powerful hardware. These developments have ushered in new opportunities for the practical realization of sophisticated control algorithms. Unlike continuous-time sliding mode control strategies, which often demand extremely high switching frequencies and were historically challenging to implement, discrete-time sliding mode control presents a more feasible and straightforward option. The finite switching frequency requirement in discrete-time sliding mode control aligns well with the capabilities of modern cost-effective digital processors, making the implementation process more accessible and efficient. This convergence of technological progress and algorithmic requirements has facilitated the adoption and integration of discrete-time sliding mode control in a variety of applications.

1.3 A Comprehensive Literature Review

In recent years, the field of control systems has experienced notable progress as researchers have investigated diverse methodologies to address the complexities presented by real-world systems, characterized by plant uncertainties and external disturbances. This literature review offers a comprehensive overview of current research in discrete-time sliding mode control, encompassing reaching law-based approaches, equivalent control-based approaches, higher-order sliding mode control, discrete-time twisting algorithm, discrete-time super-twisting algorithm, and related areas. It aims to illuminate the key contributions and emerging trends within this dynamic field.

1.3.1 Sliding Mode Control

Sliding mode control, developed in the 1950s for continuous-time systems (CTS), is a well-established and robust control technique [1–5]. Renowned for its effectiveness in handling nonlinearities, time-varying dynamics, and uncertainties, SMC has demonstrated robustness against matched disturbances and offers efficient implementation [6–14]. In numerous control engineering applications [15] where achieving desired performance is essential despite discrepancies between mathematical models and real systems, as well as external disturbances, sliding mode control proves to be a highly effective technique [4, 6, 16]. Its application extended naturally to discrete-time systems (DTS), propelled by the prevalence of digital platforms such as microcontrollers. The core idea of SMC involves selecting a sliding surface to constrain the state's motion, fundamentally shaping the dynamic behavior of the closed-loop system [4, 17]. One of the standout features of SMC is its inherent robustness against matched bounded disturbances. This means that even in the presence of matched disturbances, sliding mode control can force the system to follow the desired trajectory on the sliding surface. This robustness property makes SMC particularly appealing for applications where system dynamics are subject to external perturbations or uncertainties [4, 7].

Despite its effectiveness, sliding mode control is not without challenges. The phenomenon of chattering [9, 18], characterized by high-frequency oscillations around the sliding surface, is a well-known issue. Researchers are actively engaged in developing techniques to mitigate chattering and improve the practical applicability of sliding mode control [8]. The occurrence of chattering is primarily attributed to the idealized nature of the sliding mode. In an ideal scenario, the system state should precisely follow the sliding surface, experiencing instantaneous and abrupt control inputs to maintain this ideal motion. However, in practical systems, achieving such instantaneous changes is not feasible due to physical limitations, actuator dynamics, and the discrete nature of control implementations. Over the past years, so many methods have been developed to curb the chattering effects [19, 20].

1.3.2 Discrete-Time Sliding Mode Control

With the view of practical implementation, in this era of digital computers and microcontrollers, the discrete-time sliding mode control (DSMC) has been explored by many authors [21–32], just to name a few. The fundamental distinction between discrete-time sliding mode control and continuous-time sliding mode control (CSMC) lies in DSMC's finite switching frequency, negating the invariance property which is inherent in CSMC. In most part of the design, discrete sliding mode control has been further explored along two main directions. First, the reaching law-based DSMC which incorporates the switching term explicitly, resulting in unavoidable chattering [26, 33]. Secondly, equivalent control-based DSMC, as investigated by Utkin [23, 34] and others [35–37], which omits the reaching phase but may necessitate substantial control effort. Moreover, [28] also explored non-switching type reaching law but without the need for excessive control effort. Further, the notion of quasi-sliding mode (QSM) was proposed for discrete-time systems since the ideal sliding mode could not occur for such systems. Likewise many properties of CSMC for DTS could not propagate that well as it was for continuous-time systems. Finite-time stability (FTS) in systems employing discrete-time sliding mode control (DSMC) has historically received limited attention. However, recent years have seen advancements in this area. A minimum function based method to yield finite-time stability for discrete-time systems was studied in [38–40], demonstrating its applicability in the domain of optimization [41], where the trajectories converges to the origin in finite span of time rather than remaining within an ultimate bound. This concept has since been extended to develop discrete sliding mode control [42] and a discrete-time super-twisting-like algorithm [43].

In the reaching law approach [26–29, 44–50], the desired value of the switching variable for the next time step is determined by the reaching law, which is based on the current switching variable. This methodology finds its suitability for discrete-time systems, ensuring a guaranteed sliding mode. Originally introduced for continuous-time systems in [33], this approach was extended to DTS in [26]. To address non-ideal sliding motion in DTS, the concept of quasi-sliding mode (QSM) was introduced and further investigated in [27], incorporating additional conditions for its existence. Various approaches have been explored to mitigate chattering in discrete-time sliding mode control, as evidenced by several studies. In [28], the discussion centers around quasi-sliding mode control, with the goal of maintaining states near the switching manifold without the need for recrossing and, consequently, avoid chattering. The work in [51] introduces a robust discrete sliding mode algorithm designed to ensure chattering-free operation while effectively handling parameter uncertainties and external disturbances. The study in [52] delves into non-smooth DSMC as a strategy to address chattering induced by reaching law-based control. Furthermore, the merits of chattering-free DSMC are exemplified through a significant case study involving missile guidance systems, as demonstrated in [53].

Over the past decade, extensive research has been conducted in the realm of discrete-time sliding mode control, encompassing a range of areas[46, 54–57]. Notable contributions include investigations into discrete-time integral SMC tailored for sampled-data systems [58], the exploration of discrete-time terminal SMC [59], and its practical application in addressing perturbations in automotive electronic throttle systems [60]. The advancement of DSMC methodologies includes the development of generalized reaching law-based DSMC incorporating variable functions [47] and exploring different convergence rates for them [49, 61]. Additionally, studies have delved into band approach analysis, considering predefined band sizes [48], robustness enhancement through relative degree two switching variables [50], its generalization for arbitrary relative degrees [44], and the implementation of reference switching profile-based DSMC [45]. These diverse research endeavors contribute significantly to the evolving landscape of DSMC applications and methodologies.

1.3.3 Discrete-Time Twisting Algorithm

The twisting algorithm, similar to the super-twisting algorithm, is greatly utilized in control applications because of its simple parameter configuration and robustness against model uncertainties. Its well-posedness, stability, and key qualitative properties, including tracking precision and rate of convergence, have drawn considerable interest [19]. In comparison to the super-twisting algorithm, the numerical chattering of the twisting algorithm becomes more pronounced due to its composition of two cascaded signum functions and the absence of an integration operation. Efforts to mitigate numerical chattering have been extensively explored in the literature, encompassing both discrete-time approaches [52, 62–64] and continuous-time techniques, such as higher-order sliding mode (HOSM) control [19].

As highlighted in [65, 66], implicit-Euler methods have demonstrated significant potential in reducing numerical chattering as opposed to traditional discrete-time implementations utilizing explicit Euler methods. For instance, [67] introduced an implicit-Euler-based approach to mitigate numerical chattering in the twisting algorithm, analyzing its stability and robustness in both disturbed and disturbance free scenarios utilizing the discrete-time Lyapunov approach. Nevertheless, this approach only accounts for ideal implicit-Euler discretization, overlooking the zero-order hold (ZOH) effect caused by analog-to-digital converters in practical systems.

To address the ZOH effect, [68] discussed an implicit-Euler approach that incorporates ZOH discretization. However, this solution requires numerical solvers to resolve the affine variational inequality for computing the twisting algorithm's control input, which can be impractical for computation-constrained embedded systems and may compromise real-time performance. To overcome these limitations, [69] presented an implicit-Euler ZOH discretization that eliminates the need for numerical solvers.

It is important to note that these works primarily focus on discretizing the continuous-time twisting algorithm to facilitate its implementation using digital controllers.

1.3.4 Discrete-Time Super-Twisting Algorithm

The super-twisting algorithm (STA), introduced by Levant in 1993 [19], is a widely recognized tool for control and observation for nonlinear systems. As a second-order sliding mode algorithm, it is designed to cater plants with relative degree one in reference with specified switching function, often the variable considered to be output. A key notable feature is its ability to ensure finite-time convergence of the controlled variable even under matched Lipschitz perturbations. Unlike first-order sliding mode controllers, the STA produces a continuous control law, a significant advantage in various practical applications. Additionally, the STA is straightforward to implement in digital environments.

In most of the existing literature, discrete-time higher-order sliding mode controllers have been developed using both explicit and implicit Euler discretization methods [59, 62, 63, 66, 67, 69–74]. Among these, the explicit Euler discretization method is the most commonly used for digital implementation of the STA [37, 71, 73, 74]. When implemented with explicit Euler discretization, the closed-loop plant retains the asymptotic precision observed in the continuous-time plant.

However, studies have shown that implicit Euler-discretized STAs [63, 66, 67] outperform explicit methods in suppressing chattering. Implicit Euler discretization provides unconditional stability, enabling it to handle a broader range of problems without risking numerical instability. By contrast, explicit Euler discretization is conditionally stable and prone to numerical instabilities [66]. Additionally, implicit Euler-discretized STAs have demonstrated greater robustness to noise and uncertainties, improved accuracy with larger sampling times, and better handling of nonlinearities [67].

The implicit method also offers exact convergence in the unperturbed case, steering the controlled variable precisely to origin. Even in the presence of disturbances, it maintains the optimal asymptotic accuracy characteristic of the continuous-time system. An important benefit of the implicit methodology is that its accuracy is unaffected by increase in control gains.

Nevertheless, we clearly note that discretizing continuous-time sliding mode algorithms can sometimes lead to periodic behavior [75–77] or, in more severe cases, unstable closed-loop dynamics [78, 79].

1.4 Contributions of the Monograph

This monograph stands as a notable contribution to the expansive domain of control systems, presenting an innovative approach through the introduction and application of sliding mode based on difference equations with minima for dynamical systems. The profound impact of this research extends across various dimensions, with key contributions manifesting in the following facets:

- We begin with a thorough exploration of the history and development of discrete-time sliding mode control, grounded in the foundational principles of continuous-time sliding mode control. It first presents a basic design methodology for CSMC, followed by an overview of DSMC's evolution, emphasizing the critical differences in their respective design strategies. Additionally, we introduce a key classification that differentiates between switching and non-switching reaching law developments within these control methods. These foundational insights are vital for framing and supporting the research contributions discussed in the chapters that follow.
- Subsequently, our exploration introduced two new reaching laws specifically tailored for discrete-time plants, utilizing a difference equation with minimum

function. These reaching laws not only eliminate chattering effectively, ensuring smoother operation, but also enhance system robustness. This improvement is realized by strategically reducing the width of the quasi-sliding mode band without imposing excessive stress on the control effort. The nuanced balance achieved by these reaching laws demonstrates their efficacy in addressing challenges within discrete-time systems. The proposed methodology's effectiveness and robustness were validated through theoretical analysis and extensive simulation studies, demonstrating its ability to achieve sliding mode for the unperturbed case and quasi-sliding mode for the perturbed case.

- Next, we extended the discussion on reaching laws based on the difference equation with minima, introducing a shift to rate-regulatory function-based reaching laws. The primary objective was to leverage the benefits associated with the variable rate of change of the switching function, facilitated by the inclusion of a variable gain function in the proportional term. This modification led to a significantly accelerated rate of change of the switching function. Importantly, we demonstrated that this rate of change is effectively upper-bounded by the design parameters, introducing an additional advantageous feature. We conducted simulations on a numerical example and a pendulum system. In the case of the pendulum system, a comparative analysis among various methods revealed that the proposed method consistently outperformed others.
- Further, this monograph delved into the Lyapunov characterization of discrete-time sliding mode utilizing difference equation with minima. To achieve this, we formulated Lyapunov conditions for the finite-time input-to-state stability of perturbed discrete-time plants, shedding light on the settling time function. Moreover, we also discussed the Lyapunov stability of the closed-loop plant. Additionally, we conducted a comparative study through simulations, emphasizing the superior performance of the proposed laws compared to existing ones. An intriguing observation from this analysis is the attainment of the least ultimate bound for the sliding variable trajectory with satisfactory transient behavior.
- Additionally, our focus shifted to the development of discrete-time twisting-like and super-twisting-like algorithms, deeply rooted in the framework of a difference equation with minima. A departure from the common practice of Euler-discretization in conventional discrete-time algorithms, our proposed designs displayed notable distinctiveness. Leveraging these innovative designs, we successfully achieved the convergence of plant trajectories to the origin in the case of unperturbed systems and to an invariant set for plants affected by disturbances, all within a finite span of time. Furthermore, we expanded the application of the discrete-time super-twisting-like algorithm by employing it to design a super-twisting observer. This observer, in a noteworthy turn of events, robustly estimated unknown state variables for a perturbed pendulum system, demonstrating its efficacy within a finite timeframe.
- Finally, we aimed towards the application of the proposed reaching law utilizing difference equation with minimum function in a benchmark practical setup, delving into the intricacies of the applied reaching law. This real-time realization offers a comprehensive analysis of the reaching law's performance within the

context of the couple tank system. The results not only demonstrate the robustness of the proposed approach but also underscore its adaptability to real-world experimental conditions. We successfully implemented the proposed law in the couple tank system, presenting the outcomes through a combination of simulation and experimentation. The level control of the tank system was meticulously conducted under both unperturbed and perturbed conditions. In both scenarios, the successful attainment of the desired water level serves as a testament to the efficacy of the proposed laws.

1.5 Organization of the Monograph

This monograph comprises a total of eight chapters.

In the inaugural chapter of the monograph, Chap. 1 offers a comprehensive introduction to the conducted work. This includes a nuanced exploration of the motivation driving the research, an in-depth literature survey to discern the existing research gap, and a meticulous delineation of the principal contributions embedded within the monograph. This chapter not only establishes the foundation for the subsequent sections but also strategically defines the research objectives. Furthermore, it underscores the significance and relevance of the proposed work in the broader context of the academic and research landscape.

Chapter 2 serves as a comprehensive exploration of the mathematical preliminaries essential for grasping the intricacies presented throughout the monograph. Within this chapter, we delve into key notions, definitions, and concepts that lay the groundwork for a better understanding of the subsequent chapters. This encompassing overview includes critical topics such as comparison functions, the concepts of sliding mode and quasi-sliding mode, and stability notions framed within the discrete-time systems paradigm. By delving into these foundational mathematical elements, this chapter serves as a pivotal resource, establishing the necessary framework for engaging with the advanced concepts and methodologies expounded upon in the subsequent sections of the monograph.

Chapter 3 offers a thorough examination of the history and evolution of discrete-time sliding mode control, starting with the foundational principles of continuous-time sliding mode control. It begins by outlining a basic design methodology for CSMC, then traces the origins and development of DSMC, with particular attention to the key differences in their design approaches. The chapter also introduces an important classification that distinguishes between switching and non-switching reaching laws development within these control strategies. These foundational concepts are essential for contextualizing and supporting the research contributions presented in the subsequent chapters.

Chapter 4 embarks on a detailed exploration where we introduce two reaching laws crafted for discrete-time systems, leveraging the difference equations with minima. The definitions of sliding mode and quasi-sliding mode are provided. To validate their effectiveness and robustness, the chapter engages in theoretical analysis and conducts extensive simulation studies. The methodology's prowess is thereby confirmed, demonstrating its adeptness in achieving sliding mode for the unperturbed case and quasi-sliding mode for the perturbed case.

In Chap. 5, we delve deeper into the realm of reaching laws grounded in the difference equation with minima, expanding our exploration by introducing rate-regulatory function-based reaching laws. The desired properties of the rate-regulatory function are highlighted. To validate and showcase the efficacy of this approach, simulations were conducted on both a numerical example and a pendulum system. Moreover, in the case of the pendulum system, a comparative analysis among various methods was undertaken to observe the performance of all these methods together.

Moving on to Chap. 6, the monograph delves into the intricacies of the Lyapunov characterization of discrete sliding mode based on a difference equation with minimum function. In the pursuit of this objective, we meticulously laid out the Lyapunov conditions, unraveling the finite-time input-to-state stability of perturbed discrete-time plants, thereby shedding light on the nuances of the settling time function. Beyond theoretical exploration, we engaged in a comparative study through simulations to observe the performance of the proposed law in contrast to the existing methodologies.

Chapter 7 immersed itself in the design of a new type of discrete-time twisting-like and super-twisting-like algorithms, firmly grounded in the framework of a difference equation with minima. We provided the rigorous theoretical analysis to develop the stability conditions of the proposed algorithms by using Lyapunov approach. Additonally, we also provided simulation results to understand the behaviour of the algorithms by means of graphical illustration. Moreover, we expanded the utility of the discrete-time super-twisting-like algorithm to design a discrete super-twisting observer. This observer is implemented for the states estimation problem of a perturbed pendulum system in a finite-time framework.

Chapter 8 is dedicated to the practical application of the proposed reaching law based on difference equation with minima. This segment provides a detailed analysis of the reaching law's performance within the realm of the couple tank system. We have conducted simulation study as well as experimental study for the tank system. This study is performed for the level control of the tank system which is scrutinized under both unperturbed and perturbed conditions.

Finally, the concluding chapter encapsulates the essential discoveries and contributions outlined in the monograph, offering a comprehensive overview. It delves into the implications of the research, acknowledges its limitations, and outlines potential paths for future research and practical applications. This chapter serves as the culmination of the monograph, providing a sense of closure to the overarching study.

References

1. Emelyanov, S.V.: A method to obtain complex regulation laws using only the error signal or the regulated coordinate and its first derivatives. Avtomat. Telemekh. **18**(10), 873–885 (1957)
2. Utkin, V.I.: Variable structure systems with sliding modes. IEEE Trans. Autom. Control **AC-22**(2), 212–222 (1977)
3. Utkin, V.I.: Sliding Modes in Control and Optimization. Springer-Verlag, Berlin, Germany (1992)
4. Edwards, C., Spurgeon, S.: Sliding mode control: theory and applications. CRC Press (1998)
5. Posielek, T., Wulff, K., Reger, J.: Analysis of sliding-mode control systems with relative degree altering perturbations. Automatica **148**, 110745 (2023)
6. Young, K.D., Utkin, V.I., Ozguner, U.: A control engineer's guide to sliding mode control. IEEE Trans. Control Syst. Technol. **7**(3), 328–342 (1999)
7. Perruquetti, W., Barbot, J.P.: Sliding mode control in engineering. CRC Press (2002)
8. Laghrouche, S., Plestan, F., Glumineau, A.: Higher order sliding mode control based on integral sliding mode. Automatica **43**(3), 531–537 (2007)
9. Shtessel, Y., Edwards, C., Fridman, L., Levant, A.: Sliding mode control and observation, 1st edn. Springer, Birkhauser, New York (2014)
10. Kumari, K., Bandyopadhyay, B., Reger, J., Behera, A.K.: Event-triggered sliding mode control for a high-order system via reduced-order model based design. Automatica **121**, 109163 (2020)
11. Zhang, L., Chen, Z., Yu, X., Yang, J., Li, S.: Sliding-mode-based robust output regulation and its application in PMSM servo systems. IEEE Trans. Ind. Electr. **70**(2), 1852–1860 (2023)
12. Thomas, M., Bandyopadhyay, B.: Discrete-time sliding mode control design for unicycle robot with bounded inputs. IEEE Trans. Circuits Syst. II: Express Briefs **68**(8), 2912–2916 (2021)
13. Pal, A.K., Kamal, S., Yu, X., Nagar, S.K., Bandyopadhyay, B.: Free-will arbitrary time terminal sliding mode control. IEEE Trans. Express Briefs, Circuits Syst. II (2020)
14. Singh, B., Pal, A.K., Kamal, S., Dinh, T.N., Mazenc, F.: Vector control Lyapunov function based stabilization of nonlinear systems in predefined time. IEEE Trans, Autom Control (2022)
15. Plestan, F., Shtessel, Y., Br geault, V., Poznyak, A.: Sliding mode control with gain adaptation: Application to an electropneumatic actuator. Control Eng. Pract. **21**(5), 679–688 (2013)
16. Utkin, V.I., Vadim, I.: Sliding mode control. Variable structure systems: from principles to implementation 66 (2004)
17. Fridman, L., Moreno, J.A., Bandyopadhyay, B., Kamal, S., Chalanga, A.: Continuous nested algorithms: The fifth generation of sliding mode controllers. In: Recent advances in sliding modes: From control to intelligent mechatronics, pp. 5–35. Springer (2015)
18. Lee, H., Utkin, V.I.: Chattering suppression methods in sliding mode control systems. Ann. Rev. Control **31**(2), 179–188 (2007)
19. Levant, A.: Sliding order and sliding accuracy in sliding mode control. Int. J. Control **58**(6), 1247–1263 (1993)
20. Plestan, F., Glumineau, A., Laghrouche, S.: A new algorithm for high-order sliding mode control. Int. J. Robust Nonlinear Control **18**(4–5), 441–453 (2008)
21. Milosavljevic, D.: General conditions for the existence of a quasi-sliding mode on the switching hyperplane in discrete variable structure systems. Autom. Remote. Control. **46**(3), 307–314 (1985)
22. Sarpturk, S., Istefanopulos, S.Y., Kaynak, O.: On the stability of discrete-time sliding mode control systems. IEEE Trans. Automat. Contr. **32**, 930–932 (1987)
23. Utkin, V.I., Drakunov, S.V.: On discrete-time sliding modes. IFAC Proc. **22**(3), 273–278 (1989)
24. Furuta, K.: Sliding mode control of a discrete system. Syst. Control Lett. **14**(2), 145–152 (1990)
25. Sira-Ramirez, H.: Non-linear discrete variable structure systems in quasi-sliding mode. Inter. J. Control **54**(5), 1171–1187 (1991)
26. Gao, W., Wang, Y., Homaifa, A.: Discrete-time variable structure control systems. IEEE Trans. Ind. Electron. **42**(2), 117–122 (1995)
27. Bartoszewicz, A.: Remarks on discrete-time variable structure control systems. IEEE Trans. Ind. Electron. **43**(1), 235–238 (1996)

28. Bartoszewicz, A.: Discrete-time quasi-sliding-mode control strategies. IEEE Trans. Ind. Electron. **45**(4), 633–637 (1998)
29. Bartoszewicz, A., Lesniewski, P.: New switching and nonswitching type reaching laws for SMC of discrete time systems. IEEE Trans. Control Syst. Tech. **24**(2), 670–677 (2016)
30. Zhang, J., Shi, P., Xia, Y., Yang, H.: Discrete-time sliding mode control with disturbance rejection. IEEE Trans. Indus. Electr. **66**(10), 7967–7975 (2018)
31. Hettiger, C., Watermann, L., Kumari, K., Eisenzopf, L., Weissenberger, F., Horn, M., Koch, S., Reger, J., Reichhartinger, M.: On discretization methods for indirect adaptive sliding mode control. In: IEEE 61st Conference on Decision and Control (CDC), pp. 4930–4936 (2022)
32. Michel, N., Olaru, S., Valmorbida, G., Bertrand, S., Dumur, D.: Invariant sets for discrete-time constrained linear systems using a sliding mode approach. In: 2018 European Control Conference (ECC), pp. 2929–2934 (2018)
33. Gao, W., Hung, J.C.: Variable structure control of nonlinear systems: A new approach. IEEE Trans. Ind. Electron. **40**(1), 45–55 (1993)
34. Utkin, V.I.: Sliding mode control in discrete-time and difference systems. In: Variable Structure and Lyapunov Control, pp. 87–107 (1994)
35. Su, W.C., Drakunov, S.V., Ozguner, U.: An $O(T^2)$ boundary layer in sliding mode for sampled-data systems. IEEE Trans. Autom. Control **45**(3), 482–485 (2000)
36. Yu, X., Chen, G.: Discretization behaviors of equivalent control based sliding-mode control systems. IEEE Trans. Autom. Control **48**(9), 1641–1646 (2003)
37. Galias, Z., Yu, X.: Euler's discretization of single input sliding-mode control systems. IEEE Trans. Autom. Control **52**(9), 1726–1730 (2007)
38. Haddad, W.M., Lee, J.: Finite-time stability of discrete autonomous systems. Automatica **122** (2020)
39. Lee, J., Haddad, W.M., Bhat, S.P.: Finite time stability of discrete-time stochastic dynamical systems. In: 60th IEEE Conf. Deci. Control, pp. 6646–6651. IEEE (2021)
40. Haddad, W.M., Lee, J.: Finite-time stabilization and optimal feedback control for nonlinear discrete-time systems. IEEE Trans. Autom. Control (2022). https://doi.org/10.1109/TAC.2022.3151195
41. Prasun, P., Pandey, S., Kamal, S., Ghosh, S., Singh, D.: Discrete-time gradient systems governed by difference equation with minima. In: 31st Mediterranean Conf. Contr. and Autom, Limassol, Cyprus (2023)
42. Prasun, P., Kamal, S., Bartoszewicz, A., Ghosh, S.: Difference equation with minima based discrete-time sliding mode control. IEEE Trans. Circ. and Syst. II: Expr. Briefs **70**(12), 4404–4408 (2023)
43. Prasun, P., Pandey, S., Kamal, S., Ghosh, S., Xiong, X.: A minimum operator based discrete-time super-twisting-like algorithm. IEEE Trans. Circ. and Syst. II: Expr. Briefs **71**(1), 286–290 (2023)
44. Bartoszewicz, A., Latosiński, P.: Generalization of Gao's reaching law for higher relative degree sliding variables. IEEE Trans. Autom. Control **63**(9), 3173–3179 (2018)
45. Bartoszewicz, A., Adamiak, K.: Discrete-time sliding-mode control with a desired switching variable generator. IEEE Trans. Autom. Control **65**(4), 1807–1814 (2020)
46. Bartoszewicz, A., Latosiński, P.: Discrete time sliding mode control with reduced switching-a new reaching law approach. Int. J. Robust Nonlinear Control **26**(1), 47–68 (2016)
47. Chakrabarty, S., Bandyopadhyay, B.: A generalized reaching law for discrete time sliding mode control. Automatica **52**, 83–86 (2015)
48. Chakrabarty, S., Bandyopadhyay, B.: Minimum ultimate band design of discrete sliding mode control. Asian J. Control **17**(5), 1889–1897 (2015)
49. Chakrabarty, S., Bandyopadhyay, B.: A generalized reaching law with different convergence rates. Automatica **63**, 34–37 (2016)
50. Chakrabarty, S., Bartoszewicz, A.: Improved robustness and performance of discrete time sliding mode control systems. ISA Trans. **65**, 143–149 (2016)
51. Golo, G.: Milosavljević, Č: Robust discrete-time chattering free sliding mode control. Syst. Control Lett. **41**(1), 19–28 (2000)

52. Du, H., Yu, X., Chen, M.Z.Q., Li, S.: Chattering-free discrete-time sliding mode control. Automatica **68**, 87–91 (2016)
53. Wang, X., Wu, D., He, J., Zhang, J., Li, S.: Chattering-free discrete-time sliding mode control for integrated missile guidance and control system. Proceedings of the Institution of Mechanical Engineers, Part G: J. Aerosp. Eng. **237**(7), 1546–1556 (2023)
54. Kurode, S., Bandyopadhyaya, B., Gandhi, P.S.: Discrete sliding mode control for a class of underactuated systems. In: 37th Annu. Conf. IEEE Ind. Electron. Soc. pp. 3936–3941 (2011)
55. Sharma, N.K., Roy, S., Janardhanan, S., Kar, I.N.: Adaptive discrete-time higher order sliding mode. IEEE Trans. Circuits Syst. II: Express Briefs **66**(4), 612–616 (2018)
56. Latosinski, P., Bartoszewicz, A.: Model reference DSMC with a relative degree two switching variable. IEEE Trans. Autom. Control **66**(4), 1749–1755 (2020)
57. Mishra, J.P., Yu, X.: Delta-operator-based reaching laws for sliding mode control design. IEEE Trans. Circuits Syst. II: Express Briefs **69**(4), 2136–2140 (2021)
58. Abidi, K., Xu, J.X., Yu, X.: On the discrete-time integral sliding-mode control. IEEE Trans. Autom. Control **52**(4), 709–715 (2007)
59. Li, S., Du, H., Yu, X.: Discrete-time terminal sliding mode control systems based on Euler's discretization. IEEE Trans. Autom. Control **59**(2), 546–552 (2013)
60. Song, M., Li, S., Du, H., Wang, Z.: Chattering-free discrete-time fast terminal sliding mode control of automotive electronic throttle with disturbances. In: 39th Chinese Control Conference. Shenyang, China (2020)
61. Ren, Q., Kao, Y., Zhu, Y., Xia, H., Wang, C.: A new discrete reaching condition and generalized discrete reaching law with different convergence rates. Automatica **132** (2021)
62. Acary, V., Brogliato, B., Orlov, Y.V.: Chattering-free digital sliding-mode control with state observer and disturbance rejection. IEEE Trans. Autom. Control **57**(5), 1087–1101 (2011)
63. Xiong, X., Pal, A.K., Liu, Z., Kamal, S., Huang, R., Lou, Y.: Discrete-time adaptive super-twisting observer with predefined arbitrary convergence time. IEEE Trans. Circ. Syst. II: Expr. Briefs **68**(6), 2057–2061 (2021)
64. Yan, Y., Galias, Z., Yu, X., Sun, C.: Euler's discretization effect on a twisting algorithm based sliding mode control. Automatica **68**, 203–208 (2016)
65. Huber, O., Acary, V., Brogliato, B.: Lyapunov stability analysis of the implicit discrete-time twisting control algorithm. IEEE Trans. Autom. Control **65**(6), 2619–2626 (2020)
66. Brogliato, B., Polyakov, A., Efimov, D.: The Implicit discretization of the supertwisting sliding-mode control algorithm. IEEE Trans. Autom. Control **65**, 3707–3713 (2020)
67. Xiong, X., Kikuuwe, R., Kamal, S., Jin, S.: Implicit-Euler implementation of super-twisting observer and twisting controller for second-order systems. IEEE Trans. Circuits Systems II: Express Briefs **67**(11), 2607–2611 (2020)
68. Huber, O., Acary, V., Brogliato, B., Plestan, F.: Implicit discrete-time twisting controller without numerical chattering: Analysis and experimental results. Control Eng. Practice **46**, 129–141 (2016)
69. Xiong, X., Bai, Y., Shi, R., Kamal, S., Wang, Y., Lou, Y.: Discrete-time twisting algorithm implementation with implicit-Euler ZOH discretization method. IEEE Trans. Circuits Syst. II: Express Briefs **69**(8), 3435–3439 (2022)
70. Salgado, I., Kamal, S., Chairez, I., Bandyopadhyay, B., Fridman, L.: Super-twisting-like algorithm in discrete time nonlinear systems. In: The Inter. Conf. on Advanced Mechatronic Systems **2011**, 497–502 (2011)
71. Salgado, I., Kamal, S., Bandyopadhyay, B., Chairez, I., Fridman, L.: Control of discrete time systems based on recurrent super-twisting-like algorithm. ISA Trans. **64**, 47–55 (2016)
72. Koch, S., Reichhartinger, M.: Discrete-time equivalents of the super-twisting algorithm. Automatica **107**, 190–199 (2019)
73. Yan, Y., Yu, X., Sun, C.: Discretization behaviors of a super-twisting algorithm based sliding mode control system. Int. Works. Recent Adva. Sliding Modes (RASM), pp. 1–5 (2015)
74. Salgado, I., Chairez, I., Bandyopadhyay, B., Fridman, L., Camacho, O.: Discrete-time nonlinear state observer based on a super twisting-like algorithm. IET Contr. Theo. Appl. **8**, 803–812 (2014)

75. Xia, X., Zinober, A.S.I.: Δ-Modulated feedback in discretization of sliding mode control. Automatica **42**, 771–776 (2006)
76. Levant, A., Livne, M., Lunz, D.: On discretization of high-order sliding modes. In: Recent Trends Sliding Mode Cont., 102, 177 (2016)
77. Galias, Z., Yu, X.: Study of periodic solutions in discretized two-dimensional sliding-mode control systems. IEEE Trans. Circuits Syst. II: Express Briefs **58**(6), 381–385 (2011)
78. Janardhanan, S., Bandyopadhyay, B.: On discretization of continuous-time terminal sliding mode. IEEE Trans. Autom. Control **51**, 1532–1536 (2006)
79. Behera, A.K., Bandyopadhyay, B.: Steady-state behaviour of discretized terminal sliding mode. Automatica **54**, 176–181 (2015)

Chapter 2
Mathematical Preliminaries

Abstract This chapter provides the mathematical background necessary for the monograph, including the notations, fundamental definitions, and key lemmas that form the foundation for the contributions presented later. Specifically, we review the concepts of comparison functions, stability notions for discrete-time systems, finite-time stability, and finite-time input-to-state stability for discrete-time systems. These foundational elements are essential for understanding and presenting our research contributions.

2.1 Notations

Throughout the monograph, the notation used adheres to standard conventions. The symbol $\mathbb{R}$ represents the set of real numbers, while $\mathbb{R}_+$ denotes the set of positive real numbers, i.e., $\mathbb{R}_+ = \{z \in \mathbb{R} \mid z > 0\}$, $\mathbb{R}_{\geq 0}$ set of non-negative real numbers. Similarly, $\mathbb{R}_{>k}$ and $\mathbb{R}_{\geq k}$ represent the sets of real numbers greater than k and greater than or equal to k, respectively, where k is a real constant. The collection of positive integers including zero is denoted by $\mathbb{Z}_{\geq 0}$. The notation $\mathbb{R}^n$ denotes the n-dimensional Euclidean field. For any $p \geq 1$, the l_p-norm of a vector $y \in \mathbb{R}^n$ is defined as $\|y\|_p = \left(\sum_{i=1}^{n} |y_i|^p\right)^{\frac{1}{p}}$. When p is not explicitly mentioned, the norm is assumed to be the $2-$norm, denoted as $|y|$ which represents $\|y\|_2$. For matrices, we represent the transpose of matrix A as $A^\top$. Moreover, for a given vector $a = [a_1, a_2, \ldots, a_n] \in \mathbb{R}^n$, diag$(a)$ denotes the diagonal matrix with elements $a_i \in \mathbb{R}$, where $i = 1, 2, ..., n$. In the context of a given set S, min$\{S\}$ represents the smallest element within the set, whereas sup$\{S\}$ denotes the supremum, which is the least upper bound, over the elements of S. If S_1, S_2, and S_3 are any sets, and we have two functions: $g_1 : S_1 \to S_2$ and $g_2 : S_2 \to S_3$, then their composition, denoted as $g_2 \circ g_1 : S_1 \to S_3$, is defined as $(g_2 \circ g_1)(\cdot) = g_2(g_1(\cdot))$. An n-dimensional open ball B_δ with radius δ is defined as the set of points in n-dimensional Euclidean space that are at a distance less than δ from a fixed point z_0. Mathematically, it can be represented as $B_\delta = \{z \in \mathbb{R}^n : \|z - z_0\| < \delta\}$, where z_0 is the fixed point. The absolute value mapping is symbolized by $|\cdot|$. The transpose of a vector is represented as $(\cdot)^\top$, and $\lceil z \rceil$ stands for the ceiling function (smallest integer greater than or equal

S. Kamal and P. Prasun, *A Minimum Function-Based Discrete Sliding Mode Control: Theory and Applications*, Studies in Systems, Decision and Control 621,
https://doi.org/10.1007/978-981-95-1161-7_2

to z). $\text{sign}(z) = \frac{z}{|z|}$, for $z \neq 0$ and $\text{sign}(0) = 0$. l_∞ denotes the space of bounded functions. $\min(\cdot, \cdot)$ represents the minimum function and yields the minimum value at any given instant of time.

2.2 Prerequisite Definitions and Relevent Lemmas

To grasp the insights conveyed in this research, it is imperative for the reader to have a foundational understanding of the language and terminology commonly used in the examination of stability within discrete-time dynamical systems. The definitions and concepts elucidated in this chapter hold pivotal significance throughout the subsequent sections of this monograph. The majority of the foundational information in this segment has been adjusted and rephrased from the renowned textbook by H. K. Khalil, titled "Nonlinear Systems" [1] and Wasim Haddad, titled "Nonlinear Dynamical Systems and Control: A Lyapunov-Based Approach" [2], with appropriate citations included to acknowledge the source of this material.

2.2.1 Comparison Functions

We recall some definitions related to comparison functions that will be helpful in providing important definitions of stability.

Definition 2.1 (Class $\mathcal{K}$ Function) [1]
A function $\alpha : \mathbb{R}_{\geq 0} \to \mathbb{R}_{\geq 0}$ is class $\mathcal{K}$ function, if it is continuous, strictly increasing and $\alpha(0) = 0$. In addition, if $\lim_{s\to\infty} \alpha(s) = \infty$, then it is $\mathcal{K}_\infty$ function.

Examples of Class $\mathcal{K}$ Functions:

- Linear function:
$$\alpha(r) = k \cdot r, \quad k > 0$$
- Power function:
$$\alpha(r) = r^p, \quad p > 0$$
- Exponential function:
$$\alpha(r) = e^{kr} - 1, \quad k > 0$$
- Logarithmic function (defined on $(0, a)$):
$$\alpha(r) = \log(1 + r)$$

Figures 2.1 , 2.6 , 2.7 and 2.4 show the various examples of class $\mathcal{K}$ function showcasing the all three fundamental properties of the function.

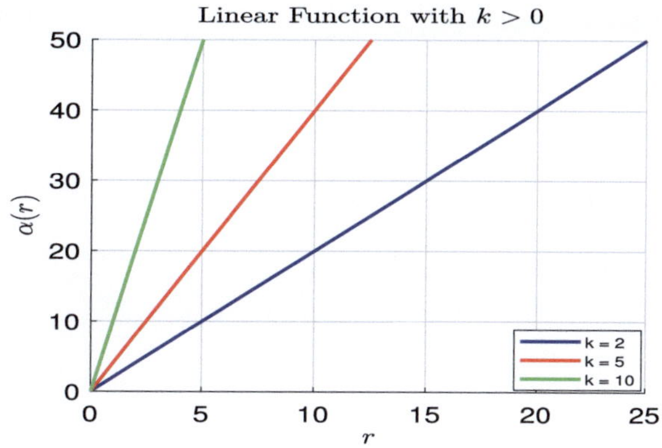

Fig. 2.1 Evolution of linear function

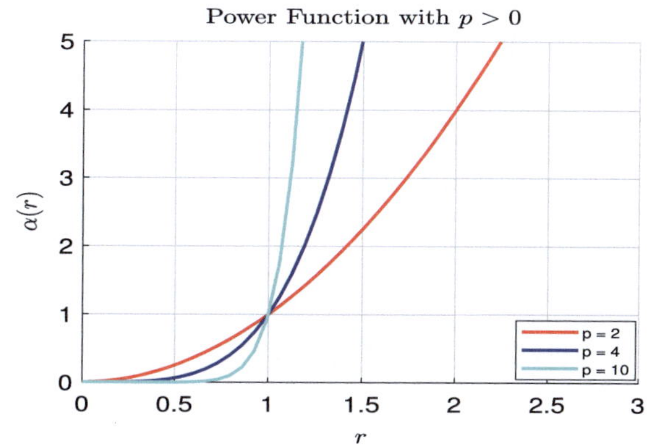

Fig. 2.2 Evolution of power function

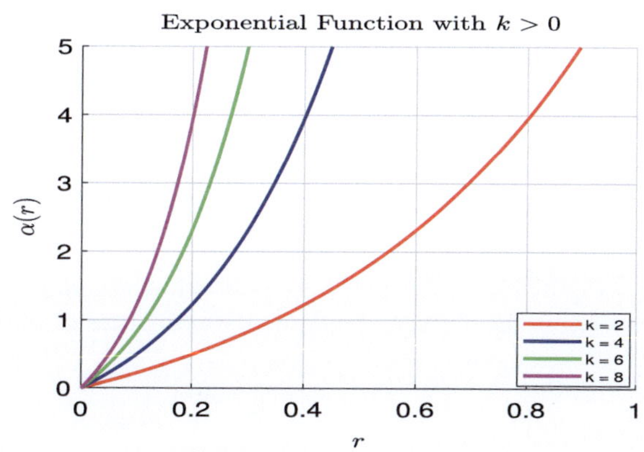

Fig. 2.3 Evolution of exponential function

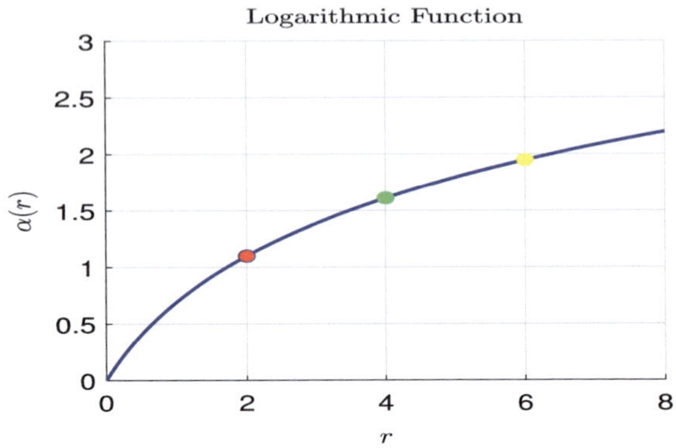

Fig. 2.4 Evolution of logarithmic function

Definition 2.2 (Class $\mathcal{L}$ Function) [1]
A function $\iota : \mathbb{R}_{\geq 0} \to \mathbb{R}_{\geq 0}$ is said to be class $\mathcal{L}$ function, if it is continuous, strictly decreasing and $\lim_{s\to\infty} \iota(s) = 0$.

Examples of Class $\mathcal{L}$ Functions:

- Exponential decay:
$$\iota(s) = e^{-ks}, \quad k > 0$$
- Inverse function:
$$\iota(s) = \frac{1}{1+s}$$
- Inverse quadratic:
$$\iota(s) = \frac{1}{1+s^2}$$
- Logarithmic decay:
$$\iota(s) = \frac{1}{\log(e+s)}$$

Figures 2.5 , 2.2 , 2.3 and 2.8 show the various examples of class $\mathcal{L}$ function depicting the all three fundamental properties of the function.

Definition 2.3 (Class $\mathcal{KL}$ Function) [1]
A function $\beta : \mathbb{R}_{\geq 0} \times \mathbb{R}_{\geq 0} \to \mathbb{R}_{\geq 0}$ is a class $\mathcal{KL}$ function if it is a class $\mathcal{K}$ function in its first argument and a class $\mathcal{L}$ function in its second argument.

Remark 2.1 A class $\mathcal{KL}$ function inherently merges the properties of class $\mathcal{K}$ and class $\mathcal{L}$ functions, as demonstrated in the following examples.

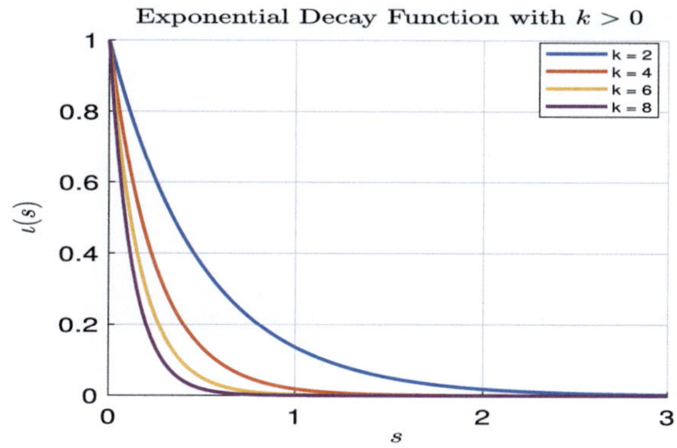

Fig. 2.5 Evolution of exponential decay function

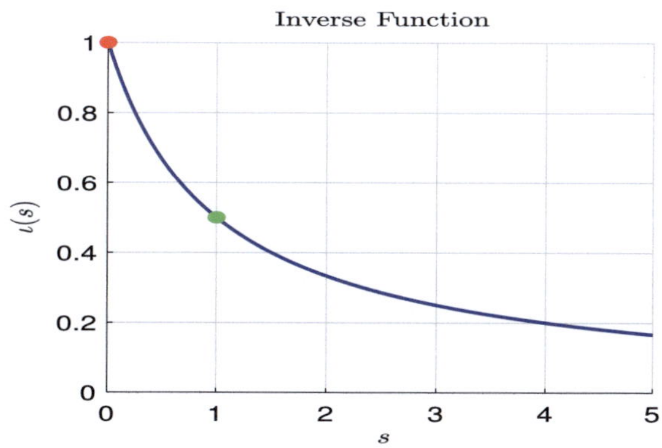

Fig. 2.6 Evolution of inverse function

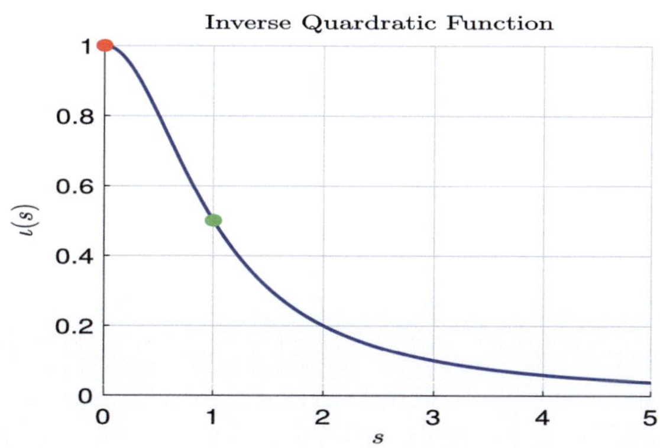

Fig. 2.7 Evolution of inverse quadratic function

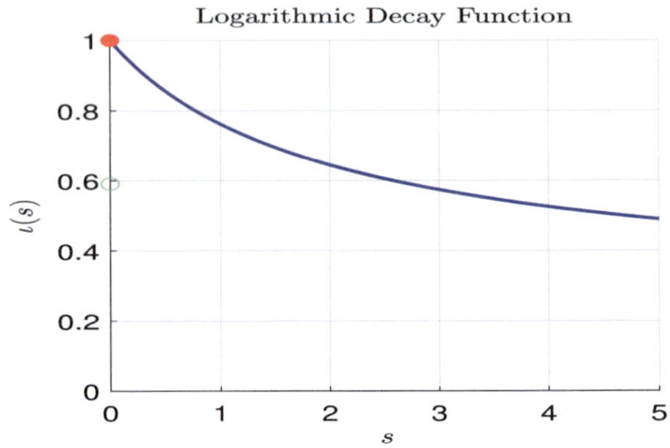

Fig. 2.8 Evolution of logarithmic decay function

Examples of Class $\mathcal{KL}$ Functions:

- Linear & exponential decay:

$$\beta(r, s) = re^{-ks}, \quad k > 0$$

- Quadratic & inverse linear decay:

$$\beta(r, s) = r^2 \cdot \frac{1}{1+s}$$

- Exponential growth & inverse quadratic decay:

$$\beta(r, s) = (e^r - 1) \cdot \frac{1}{1+s^2}$$

The following lemma outlines the interactions between these class functions and examines the preservation of class properties under inverse operations.

Lemma 2.1 *[1] Let α_1 and α_2 be class $\mathcal{K}$ functions on* $[0, a)$, *α_3 and α_4 be a class $\mathcal{K}_\infty$ functions. Denote the inverse of α_i by α_i^{-1}. Then,*

- *α_1^{-1} is defined on* $[0, \alpha_1(a))$ *and belongs to class $\mathcal{K}$.*
- *α_3^{-1} is defined on* $[0, \infty)$ *and belongs to class $\mathcal{K}_\infty$.*
- *$\alpha_1 \circ \alpha_2$ belongs to class $\mathcal{K}$.*
- *$\alpha_3 \circ \alpha_4$ belongs to class $\mathcal{K}_\infty$.*

Definition 2.4 (Generalized-$\mathcal{K}$ Function) [3]
A function $\alpha : [0, a) \rightarrow \mathbb{R}_{\geq 0}$ is called a generalized-$\mathcal{K}$ function if it follows continuity, is also increasing strictly, satisfies $\alpha(0) = 0$, and fulfills the following conditions:

$$\begin{cases} \alpha(s_1) > \alpha(s_2), \text{if } \alpha(s_1) > 0, s_1 > s_2 \\ \alpha(s_1) = \alpha(s_2), \text{if } \alpha(s_1) = 0, s_1 > s_2. \end{cases}$$

Definition 2.5 (Generalized-$\mathcal{KL}$ function) [3]
A function $\Phi : \mathbb{R}_{\geq 0} \times \mathbb{R}_{\geq 0} \to \mathbb{R}_{\geq 0}$ is called a generalized-$\mathcal{KL}$ function (or $\mathcal{GKL}$-function) if, for each fixed $t \geq 0$, the map $s \mapsto \Phi(s, t)$ is a generalized $\mathcal{K}$-map, and for every fixed $s \geq 0$, the map $t \mapsto \Phi(s, t)$ follows continuity and approaches to origin as $t \to T$, where $T \leq \infty$.

Remark 2.2 The examples presented above are consistent with the defining properties of the respective generalized class $\mathcal{K}$ and class $\mathcal{KL}$ functions.

2.3 Basic Stability Notions for Discrete-Time Systems

In this section, we revisit fundamental stability notions and definitions pertaining to the discrete-time dynamical systems.

Let us consider the following dynamical system

$$z(k+1) = f(z(k)), \ z(0) = z_0 \tag{2.1}$$

where $z \in D \subseteq \mathbb{R}^n$ represents the plant state vector, $k \in \mathbb{Z}_{\geq 0}$, $f : D \to D$, and $f(0) = 0$. It is assumed that $f(\cdot)$ is continuous on D.

Definition 2.6 [2] The zero solution $z(k) \equiv 0$ to (2.1) is Lyapunov stable if for all $\epsilon > 0$ there exists $\delta = \delta(\epsilon) > 0$ such that if $\|z(0)\| < \delta$, then $\|z(0)\| < \epsilon$, $k \in \mathbb{Z}_{\geq 0}$.

Definition 2.7 [2] The zero solution $z(k) \equiv 0$ to (2.1) is asymptotically stable if it is Lyapunov stable and there exists $\delta > 0$ such that if $\|z(0)\| < \delta$, then $\lim_{k\to\infty} z(k) = 0$.

Definition 2.8 [2] The zero solution $z(k) \equiv 0$ to (2.1) is globally asymptotically stable if it is Lyapunov stable and for all $z(0) \in \mathbb{R}^n$, $\lim_{k\to\infty} z(k) = 0$.

Definition 2.9 [2] The solution $z(k) \equiv 0$ to (2.1) is geometrically stable if positive scalars α, $\beta > 1$ exist, and another constant δ exists, such that if $\|z(0)\| < \delta$, then $\|z(k)\| \leq \alpha\|z(0)\|\beta^{-k}$, $k \in \mathbb{Z}_{\geq 0}$.

Definition 2.10 [2] The solution $z(k) \equiv 0$ to (2.1) is globally geometrically stable if positive scalars α, $\beta > 1$ exist, such that for all $z(0) \in \mathbb{R}^n$, $\|z(k)\| \leq \alpha\|z(0)\|\beta^{-k}$, $k \in \mathbb{Z}_{\geq 0}$.

Finally, the zero solution $z(k) \equiv 0$ to (2.1) is unstable if it is not Lyapunov stable.

2.4 Finite-Time Stability of Discrete-Time Systems

Let us consider the following dynamical system

$$z(k+1) = f(z(k)),\ z(0) = z_0 \tag{2.2}$$

where $z \in D \subseteq \mathbb{R}^n$ represents the plant state vector, $k \in \mathbb{Z}_{\geq 0}$, $f : D \to D$, and $f(0) = 0$.

Definition 2.11 (Finite-Time Stability) [4]
Let us consider the nonlinear plant (2.2). The solution $z(k) \equiv 0$ to (2.2) is finite-time stable if an open neighborhood $N \subseteq D$ of the equilibrium point exist and a mapping $\mathcal{K} : N\backslash\{0\} \to \mathbb{Z}_+$, called the settling time function, such that the following satisfies:

1. *Finite-time convergence:* For each $z \in N\backslash\{0\}$ and $k \geq \mathcal{K}(z)$, $z(k)$ is contained in $N \cap \{0\}$.
2. *Lyapunov stability:* For each $\epsilon > 0$, there exists $\delta > 0$ such that $B_\delta(0) \subset N$ and for each $z \in B_\delta(0)\backslash\{0\}$, $z(k) \in B_\epsilon(0)$ for every $k \in \{0, \ldots, \mathcal{K}(z_0) - 1\}$.

The trajectory $z(k) \equiv 0$ to (2.2) is globally finite-time stable if it is finite-time stable with $N = D = \mathbb{R}^n$.

The following lemma is proposed in [4] to characterize the finite-time stability.

Lemma 2.2 *[4] Let us consider the plant* (2.2). *Let there exists a continuous function $V : D \to \mathbb{R}_{\geq 0}$, which follows:*

$$V(z) > 0, \forall\, z \in D\backslash\{0\}, \tag{2.3}$$
$$V(z = 0) = 0 \text{ and} \tag{2.4}$$
$$\Delta V(z) \leq -\min(V(z), l), \forall\, z \in D\backslash\{0\} \tag{2.5}$$

where $l \in \mathbb{R}_+$. Then the trajectory $z(k) \equiv 0$ to (2.2) *exhibits finite-time stability. Additionally, there exist an open neighborhood N of the equilibrium point and a settling-time map $\mathcal{K} : N \to \mathbb{Z}_+$ such that $\mathcal{K}(z_0) \leq \left\lceil \frac{V(z_0)}{l} \right\rceil$, $z_0 \in N$, where $\mathcal{K}(\cdot)$ is lower semicontinuous on N. Moreover, if, $N = D = \mathbb{R}^n$, $V(\cdot)$ is radially unbounded, and* (2.5) *satisfies on $\mathbb{R}^n$, then the trajectory $z(k) \equiv 0$ to* (2.2) *is globally finite-time stable.*

Similarly, another characterization of finite-time stability of plant (2.2) is provided below.

Lemma 2.3 *[4] Let us take into account the plant* (2.2). *Let there exists a continuous function $V : D \to \mathbb{R}_{\geq 0}$ that follows:*

$$V(z) > 0, \forall\, z \in D\backslash\{0\}, \tag{2.6}$$
$$V(z = 0) = 0 \text{ and} \tag{2.7}$$

$$\Delta V(z) \leq -l \min\left(\frac{V(z)}{l}, (V(z))^{\sigma}\right), \forall\, z \in D\backslash\{0\} \tag{2.8}$$

where $l \in \mathbb{R}_{+}$ *and* $\sigma \in (0, 1)$. *Then the trajectory* $z(k) \equiv 0$ *to* (2.2) *exhibits finite-time stability. Additionally, there exist an open neighborhood* N *of the equilibrium point and a settling-time map* $\mathcal{K} : N \to \mathbb{Z}_{+}$ *such that* $\mathcal{K}(z_0) \leq \left\lceil \log_{[1-lV(z_0)^{\sigma-1}]} \frac{l^{\frac{1}{1-\sigma}}}{V(z_0)} \right\rceil + 1,\ z_0 \in D$, *for* $V(z_0) > l^{\frac{1}{1-\sigma}}$ *or,* $\mathcal{K}(z_0) = 1,\ z_0 \in D\backslash\{0\}$, *for* $V(z_0) \leq l^{\frac{1}{1-\sigma}}$, *where* $\mathcal{K}(\cdot)$ *is lower semicontinuous on* N. *If, in addition,* $N = D = \mathbb{R}^n$, $V(\cdot)$ *is radially unbounded, and* (2.8) *satisfies on* $\mathbb{R}^n$, *then the trajectory* $z(k) \equiv 0$ *to* (2.2) *is globally finite-time stable.*

2.5 Finite-Time Input-to-State Stability of Discrete-Time Systems

We now discuss the concept of finite-time input-to-state stability (FT-ISS) for discrete-time systems (see [5] for more details), which is useful for developing Lyapunov stability conditions for the algorithms presented in the subsequent chapters, particularly when discrete-time systems are affected by matched-type bounded perturbations (MTBP). Let us take into account the following plant:

$$z(k+1) = \psi(z(k), \nu(k)), \tag{2.9}$$

where $z \in D \subseteq \mathbb{R}^n$ represents the state vector of the system, $k \in \mathbb{Z}_{\geq 0}$, and $\nu(k) \in \mathbb{R}^m$ denotes the external input to the plant. The function $\psi : \mathbb{R}^n \times \mathbb{R}^m \to \mathbb{R}^n$ possess continuity on its domain, and follows $\psi(0, 0) = 0$. Consider the solution to (2.9) as $z(k) := z(k, z_0, \nu(k))$ for initial condition z_0.

Definition 2.12 (Finite-Time Input-to-State Stability) [5]
System (2.9) is said to be finite-time input-to-state stable if there exist $\beta(\cdot,\cdot) \in \mathcal{GKL}$ and $\gamma(\cdot) \in \mathcal{K}_\infty$ such that, for every z_0 and $\nu(k)$, the following holds: $\|z(k)\| \leq \beta(\|z_0\|, k - k_0) + \gamma(\|\nu\|)$, with $\beta(\zeta, k) = 0$ for $k \geq \mathcal{K}(\zeta)$, where $\mathcal{K}(\zeta)$ follows continuity on its argument with $\mathcal{K}(0) = 0$.

Lemma 2.4 *[5] Let us take into consideration the plant* (2.9). *Let a continuous map exists* $V : \mathbb{Z}_{\geq 0} \times D \to \mathbb{R}_{\geq 0}$, *which satisfies:*

$$\alpha_1(\|z(k)\|) \leq V(k, z) \leq \alpha_2(\|z(k)\|), \forall\, k \in \mathbb{Z}_{\geq 0}, \forall\, z \in D\backslash\{0\}, \tag{2.10}$$

$$V(k, 0) = 0, \forall\, k\ \in \mathbb{Z}_{\geq 0} \text{ and} \tag{2.11}$$

$$\Delta V(k, z) \leq -\min(V(k, z), (V(k, z))^{a}) + \alpha_3(|\nu(k)|), \forall\, k \in \mathbb{Z}_{\geq 0}, \forall\, z \in D\backslash\{0\} \tag{2.12}$$

where $a \in (0, 1)$, α_1, α_2 *and* $\alpha_3 \in$ *class* $\mathcal{K}$ *functions. Then the plant* (2.9) *is finite-time input-to-state stable. Moreover, if* $D = \mathbb{R}^2$, α_1 *and* $\alpha_2 \in$ *class* $\mathcal{K}_\infty$ *functions and* $\Delta V(k,z) \leq -\min(V(k,z),(V(k,z))^a) + \alpha_3(|\nu(k)|), \forall k \in \mathbb{Z}_{\geq 0}, \forall z \in \mathbb{R}^2 \backslash \{0\}$, *then the plant* (2.9) *is global finite-time input-to-state stable.*

2.6 Conclusion

This chapter laid the groundwork for the material presented in the following chapters. We introduced comparison functions, defined them with examples, and explored stability concepts for discrete-time systems. Additionally, we discussed finite-time stability theory for discrete-time systems, including the relevant supporting lemmas.

These foundational topics are essential, as they form the basis for the analyses and developments presented in the subsequent contributory chapters.

References

1. Khalil, H.K.: Nonlinear systems, 3rd edn. Prentice-Hall, Upper Saddle River, NJ (2002)
2. Haddad, W.M., Chellaboina, V.: Nonlinear dynamical systems and control: a Lyapunov-based approach. Princeton University Press (2008)
3. Hong, Y., Jiang, Z., Feng, G.: Finite-time input-to-state stability and applications to finite-time control design. SIAM J. Control. Optim. **48**(7), 4395–4418 (2010)
4. Haddad, W.M., Lee, J.: Finite-time stability of discrete autonomous systems. Automatica **122** (2020)
5. Liang, S., Liang, J.: Finite-time input-to-state stability of nonlinear systems: The discrete-time case. Int. J. Syst. Sci. 1–11 (2022)

Chapter 3
Continuous-Time Sliding Mode Control: Foundations and Transition to Digital Implementation

Abstract This chapter offers an in-depth exploration of the history and development of discrete-time sliding mode control (DSMC), beginning with the foundational concepts of continuous-time sliding mode control (CSMC). It first introduces a basic design approach associated with CSMC. The chapter then discusses the origin and progression of DSMC, emphasizing key distinctions between their respective design criteria. A significant classification is also presented, highlighting the switching and non-switching characteristics of these methods. These foundational insights are crucial for understanding and framing our research contributions in the later chapters.

3.1 Introduction

Sliding Mode Control (SMC) [1–4] is a nonlinear control method that provides strong robustness properties against matched uncertainties and disturbances. The key idea of SMC is to drive the system's trajectories onto a lower-dimensional manifold (called the sliding surface), and to design the control such that the system's states remain on this surface for all subsequent time. Once the sliding surface is reached, the closed-loop dynamics are insensitive to certain classes of uncertainties.

This chapter presents a self-contained treatment of conventional sliding mode control for second-order systems, complete with derivation, analysis, and an example along with the issues regarding practical implementation starting from Sects. 3.1–3.5. Later sections from Sects. 3.6–3.9 focus on inception and historical evolution of the discrete-time sliding mode control with different reaching criteria and reaching laws by the pioneers in this domain. Their limitations is also highlighted for the readers to have the understanding of why further exploration is required and to cater this the authors have build the problem statement on the basis of it.

S. Kamal and P. Prasun, *A Minimum Function-Based Discrete Sliding Mode Control: Theory and Applications*, Studies in Systems, Decision and Control 621,
https://doi.org/10.1007/978-981-95-1161-7_3

3.2 System Model and Problem Statement

Consider a second-order single-input single-output (SISO) nonlinear system of the form:

$$\ddot{x} = f(x, \dot{x}) + b(x, \dot{x})u + d(t), \tag{3.1}$$

where

- $x \in \mathbb{R}$ is the system state,
- $u \in \mathbb{R}$ is the control input,
- $f(x, \dot{x})$ is a known (or partially known) nonlinear function,
- $b(x, \dot{x}) \neq 0$ is a known control effectiveness function,
- $d(t)$ is an unknown but bounded disturbance with $|d(t)| \leq d_{\max}$.

Our objective is to design a control law $u(t)$ such that the system state $x(t)$ tracks a desired trajectory $x_d(t)$, despite the presence of the disturbance $d(t)$.

3.2.1 *Sliding Surface Design*

To design a sliding mode control law, the first step is to define the sliding surface [1, 3, 4]. This surface plays a crucial role in ensuring one of the key features of sliding mode control i.e., insensitivity to certain system uncertainties and disturbances, provided the system states remain on the surface. Therefore, we begin by focusing on the design of the sliding surface.

Let us define the tracking error as:

$$e(t) = x(t) - x_d(t), \quad \dot{e}(t) = \dot{x}(t) - \dot{x}_d(t). \tag{3.2}$$

To design an SMC law, we define a sliding surface $s(t)$ as:

$$s(t) = \dot{e}(t) + \lambda e(t) = \dot{x}(t) - \dot{x}_d(t) + \lambda(x(t) - x_d(t)), \tag{3.3}$$

where $\lambda > 0$ is a design parameter that determines the convergence rate.

When the system is on the sliding surface ($s = 0$), the tracking error dynamics reduce to:

$$\dot{e} + \lambda e = 0 \quad \Rightarrow \quad e(t) = e(0)e^{-\lambda t}, \tag{3.4}$$

which guarantees exponential convergence of $x(t)$ to $x_d(t)$.

3.2.2 Control Law Derivation

In this section, we discuss the procedure for computing the control law. To go about this problem, we differentiate the sliding variable:

$$\dot{s} = \ddot{e} + \lambda \dot{e} = \ddot{x} - \ddot{x}_d + \lambda(\dot{x} - \dot{x}_d). \tag{3.5}$$

Substituting from the system dynamics (3.1):

$$\dot{s} = f(x, \dot{x}) + b(x, \dot{x})u + d(t) - \ddot{x}_d + \lambda \dot{e}. \tag{3.6}$$

To enforce a sliding motion [5], we choose a Lyapunov function:

$$V(s) = \frac{1}{2}s^2, \tag{3.7}$$

with derivative:

$$\dot{V} = s\dot{s}. \tag{3.8}$$

To guarantee finite-time convergence [1] of $s \to 0$, enforce:

$$\dot{V} = s\dot{s} \leq -\eta |s|, \quad \eta > 0. \tag{3.9}$$

A conventional SMC law satisfying this condition is:

$$u = \frac{1}{b(x, \dot{x})} \left[-f(x, \dot{x}) + \ddot{x}_d - \lambda \dot{e} - K \,\mathrm{sign}(s) \right], \tag{3.10}$$

where $K > d_{\max}$ ensures the sliding condition is met.

3.3 Example: SMC for a Perturbed Double Integrator

Consider the perturbed second-order system:

$$\ddot{x} = u + d(t), \quad |d(t)| \leq 1. \tag{3.11}$$

Let the desired trajectory be $x_d(t) = \sin(t)$. Then:

$$\dot{x}_d(t) = \cos(t), \quad \ddot{x}_d(t) = -\sin(t).$$

We choose the sliding surface design parameter $\lambda = 2$, and define:

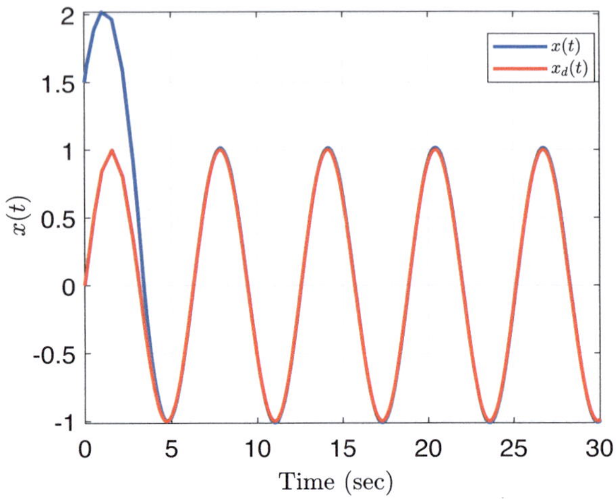

Fig. 3.1 Evolution of system state showcasing the tracking of reference signal in presence of disturbance

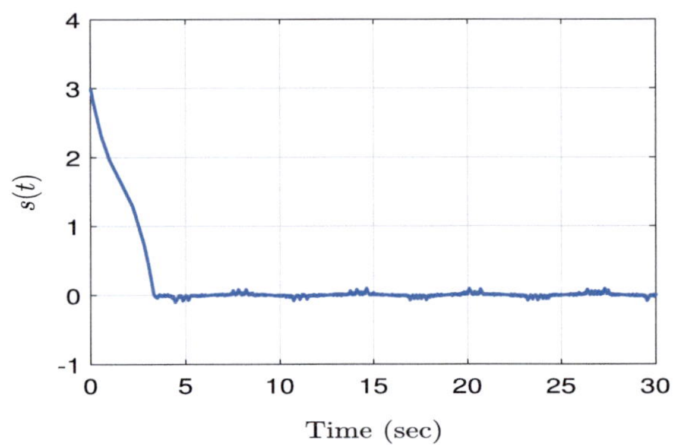

Fig. 3.2 Evolution of sliding variable depicting the finite-time convergence

$$e(t) = x(t) - \sin(t), \tag{3.12}$$
$$\dot{e}(t) = \dot{x}(t) - \cos(t), \tag{3.13}$$
$$s(t) = \dot{e}(t) + 2e(t). \tag{3.14}$$

Applying the SMC law as:

$$u(t) = -\sin(t) - 2(\dot{x}(t) - \cos(t)) - K\,\text{sign}(s(t)), \tag{3.15}$$

with $K = 1.5 > d_{\max}$. This guarantees that the sliding condition $\dot{V} < 0$ holds, and the tracking error converges exponentially to zero. For simulation purpose, we have considered $d(t) = sin(t)$. Figures 3.1, 3.2 and 3.3 present the system state, sliding variable and control signal respectively. We can see that system state perfectly tracks the reference signal even in the presence of disturbance signal. The sliding variable goes to zero in some finite time showcasing the finite-time convergence property.

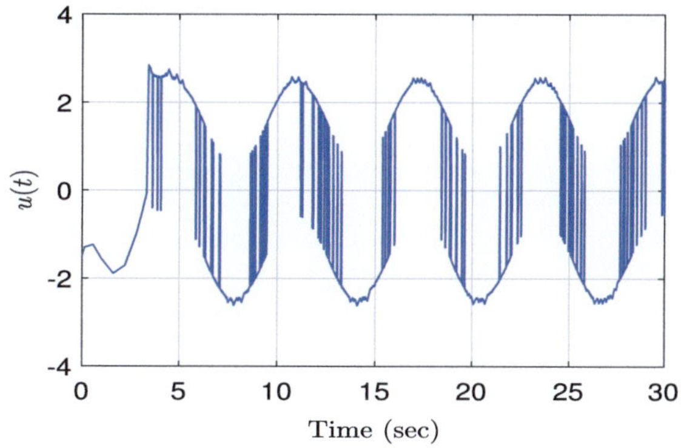

Fig. 3.3 Evolution of control signal showcasing its discontinuous nature

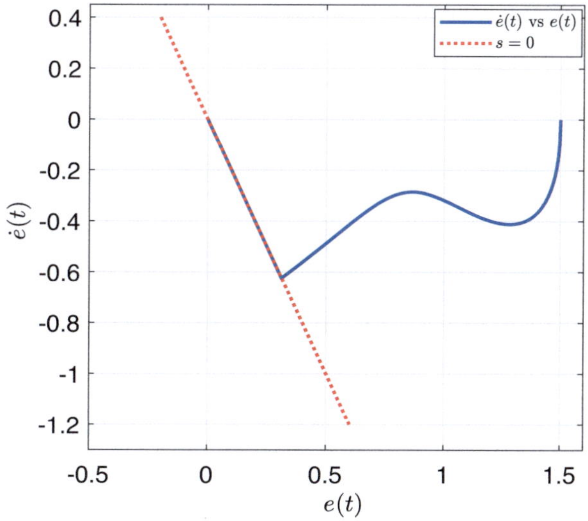

Fig. 3.4 Phase portrait of the error dynamics

Additionally, we observe that to achieve the trajectory tracking objective, the SMC control signal has discontinuous behaviour. For electro-mechanical systems, such type of control can be detrimental in nature. To mitigate this problem literature suggest to use higher-order SMC. Figure 3.4 displays the phase portrait of the error dynamical system. We can see that from some initial conditions, the error variables converge to the sliding surface in finite time. After reaching on the surface, the error variables converge to zero asymptotically. Moreover, Fig. 3.5 shows that by using SMC technique we can estimate the disturbance signal in finite amount of time. This remarkable benefit is highlighted in the following remark.

Remark 3.1 One of the key advantages of SMC is that it provides the disturbance signal estimation in finite-time, consequently helping in completely rejecting it.

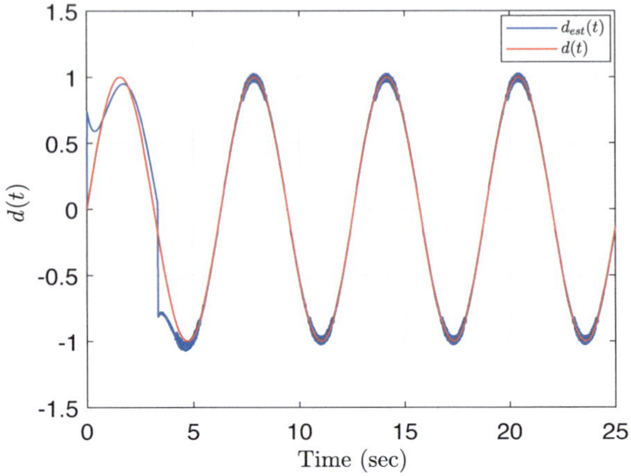

Fig. 3.5 Construction of disturbance signal in finite time

3.4 Practical Considerations and Remarks

3.4.1 Chattering

Chattering [1, 3, 4] in sliding mode control refers to the high-frequency oscillations that occur in the control signal or system states when the system operates near or on the sliding surface. It's a well-known drawback of classical SMC and can pose serious practical issues, especially for mechanical or electrical systems. The primary causes of the chattering are:

- The discontinuous nature of the $\text{sign}(s)$ function causes chattering.
- In real-world systems, switching can't happen infinitely fast due to actuator limitations or sampling rates.
- Model uncertainties and external disturbances can make it harder to stay exactly on the sliding surface, triggering frequent switching.

This can be mitigated using a boundary layer method:

$$\text{sign}(s) \to \text{sat}\left(\frac{s}{\phi}\right), \quad \phi > 0,$$

where $\text{sat}(\cdot)$ is the saturation function and ϕ defines the boundary layer thickness. Additionally, higher-order sliding mode control (HOSM) are also there which takes care of chattering problem.

3.4.2 *Robustness*

SMC is robust to matched and bounded type uncertainties (those entering the system in the same channel as control input). It refers to the controller's ability to maintain system performance in the presence of model uncertainties, parameter variations, and external disturbances.

Key Points:

- Once the system reaches the sliding surface, its motion is governed by a reduced-order sliding mode dynamics. These dynamics are insensitive to certain matched uncertainties and disturbances (i.e., those that enter the system through the same channel as the control input).
- The switching nature of SMC enables it to actively reject disturbances and compensate uncertainties. The controller adjusts rapidly to deviations from the sliding surface, forcing the trajectory back despite disturbances.

Robustness comes from the high-gain switching control, but this can also cause chattering. Therefore, there is always a trade-off between robustness and chattering.

Remark 3.2 The sliding mode control technique takes good care of matched type of disturbances. However, it does not inherently compensate for unmatched disturbances.

3.4.3 *Tuning Parameters*

Tuning parameters in sliding mode control is a critical step to ensure desired performance such as fast convergence, low chattering, and strong robustness. The main tuning parameters typically influence the sliding surface and the control law.

- λ affects the speed of convergence during sliding motion.
- K must be chosen to exceed the maximum possible disturbance magnitude $d_{\max}$.

3.5 Remarks on Conventional SMC

Conventional Sliding Mode Control offers an effective framework for robust control of uncertain systems. Through proper choice of the sliding surface and discontinuous control action, the method ensures finite-time convergence and robustness to matched disturbances. Despite the challenge of chattering, SMC remains a valuable tool in the control engineer's toolkit. As far as implementation is concerned, there is a shift towards the discrete sliding mode control because of its feasibility and compatibility with digital controllers. Next few sections focus on this very idea and captures the inception, evolution and essence of the discrete sliding mode control.

3.6 Discrete Sliding Mode Control: Inception and Evolution

Discrete-time sliding mode control emerged as a digital adaptation of continuous-time sliding mode control, aiming to maintain robustness in systems with digital implementations. DSMC is often preferred for modern control systems due to its compatibility with digital platforms such as microcontrollers, digital signal processors (DSPs), and field programmable gate arrays (FPGAs). As most real-world control systems are implemented in a discrete-time domain where signals are sampled at regular intervals, DSMC provides a robust and efficient solution for achieving desired performance. Unlike continuous systems, where real-time control is continuous, DSMC offers a practical approach to implement sliding mode control in discrete systems while addressing challenges such as sampling effects, quantization, and finite resolution of the control signal. Additionally, DSMC can be more computationally feasible in digital implementations, as control laws are updated at discrete time steps, reducing the need for continuous monitoring and adjustments. Furthermore, DSMC can be designed to reduce chattering, which is a common issue in traditional continuous sliding mode systems, by using non-switching approaches. Overall, DSMC is ideal for practical, digital applications where precision and computational efficiency are necessary, making it highly suitable for a wide range of modern systems like robotics, power electronics, and automotive control systems.

Discrete-time sliding mode control has been defined in various ways [6–12]. Milosavljevic [7] explored DSMC for discrete-time systems within the context of sampled-data systems, referring to the discrete sliding mode as the quasi-sliding mode. Sarpturk et al. [8] introduced a new sufficient condition for the existence of DSMC and examined its stability. A key challenge in discrete sliding mode control is determining an appropriate reaching condition so that, as the sample period approaches zero, the continuous-time sliding mode reaching condition is satisfied. Some researchers have applied ideal sliding mode conditions in control design Utkin [13], while others have proposed sliding mode reaching conditions for SISO systems [8, 10]. In [7], a sliding mode reaching condition is presented, but it serves only as a necessary condition for the existence of the sliding mode.

3.6.1 Evolution of DSMC and Its Existence Criteria

Analogous to the evolution of the sliding variable in continuous-time SMC, various DSMC methods have proposed existence conditions based on the evolution of the sliding variable. Key contributions in this area are summarized below.

Reaching Condition by Dote and Hoft citech3Dote1980**:**
Dote and Hoft gave the following condition which was analogous to the continuous-time counterpart. This was given as:

$$\Delta s(k)s(k) < 0 \tag{3.16}$$

where $\Delta s(k) = s(k+1) - s(k)$.
However, Milosavljevic [7] noted that because of the discretization with non-zero sampling interval, an ideal sliding mode can not exist and discussed the notion of quasi-sliding mode. For this quasi-sliding mode, only necessary conditions was derived and similar works were presented by [8, 10].

Reaching Condition by Sarpturk [8]**:**
Sarpturk et al. gave the following condition based on the absolute value of the sliding variable as follows:

$$|s(k+1)| < |s(k)| \tag{3.17}$$

Reaching Condition by Furuta [9]**:**
Furuta considered a Lyapunov function as $V(k) = 0.5s^2(k)$, made use of the discrete equivalent of the condition $\dot{V} < 0$ and deduced the following:

$$|s(k)\Delta s(k)| < -0.5(\Delta s(k))^2 \tag{3.18}$$

Reaching Condition by Ramirez [10]**:**
Ramirez proposed the following necessary reaching condition:

$$|s(k+1)s(k)| < (s(k))^2 \tag{3.19}$$

In the subsequent years, Gao et. al [11] addressed this issue of incompleteness of the reaching criteria given by the various authors for the appropriate definition of quasi-sliding mode in discrete-time systems.

3.7 Discretization of Reaching Law Approach for Continuous-Time SMC Systems

Earlier methods followed the approach of first deriving an analytical expression for the reaching condition, and then designing the control law that satisfies this condition. A different approach was devised by Gao et al. [14] called as reaching law approach. Following subsection focuses on the basic notion and development of the Gao's reaching law [11].

3.7.1 Gao's Reaching Law

According to this type of philosophy, three important attributes were listed which needed to be followed by a designed reaching law. These attributes are:

- Beginning from any initial state, the system trajectory will move steadily toward the switching surface and intersect it within a finite time.
- After the trajectory crosses the switching plane for the first time, it will continue to intersect the plane at each subsequent sampling instant, producing a zigzag motion around the switching surface.
- The magnitude of each successive zigzag step does not increase, and the trajectory remains confined within a defined boundary band.

Considering a continuous-time system, following reaching law was provided in [14] as:

$$\dot{s} = -qs - \epsilon \text{sign}(s) \tag{3.20}$$

where $q, \epsilon \in \mathbb{R}_+$ are parameters chosen by the designer.

Considering forward-Euler discretization,

$$\dot{s} = \frac{s(k+1) - s(k)}{T_s}$$

where $T_s \in \mathbb{R}_+$ is the sampling time interval.

Consequnetly, the discrete form of the continuous-time reaching law (3.20) becomes

$$s(k+1) = (1 - qT_s)s(k) - \epsilon T_s \text{sign}(s(k)) \tag{3.21}$$

where $(1 - qT_s) > 0$. For the perturbed system, the reaching law holds an extra term and presented as:

$$s(k+1) = (1 - qT_s)s(k) - \epsilon T_s \text{sign}(s(k)) + \delta(k) \tag{3.22}$$

where $\delta(k)$ is a bounded disturbance and satisfies $|\delta(k)| \leq \delta_m$.

Remark 3.3 To follow the first attribute of the reaching law mentioned above, the inequality on sampling time interval must follow. This clearly puts the constraint on the selection of the sampling time noting its one of the limitations.

Remark 3.4 The size of the ultimate band depends on the upper bound of the disturbance δ_m and the choice of the design parameter ϵ. One can think of reducing the size of this band by taking ϵ as small as possible but it would cost us in terms of reduced convergence speed.

Remark 3.5 The aforementioned strategies are mostly based on the switching function making these a replica of continuous-time laws. In the upcoming discussion, we show some other methods which do not utilize the switching function explicitly.

3.8 Non-Switching Discrete Sliding Mode Methods

Non-switching approaches emerged from the understanding that, due to the inherently discrete nature of discrete-time systems, incorporating an explicit switching function offers no significant advantage. The initial investigations into these methods were conducted by [12, 15], though each represented a distinct approach.

3.8.1 Utkin's Reaching Law

Drakunov and Utkin [15] introduced a non-switching reaching law in which the sliding variable reaches the sliding surface $s(k) = 0$ in a single time step, as opposed to the finite-time convergence proposed in [11]. This approach is inspired by the concept of deadbeat control in discrete-time systems, where the steady-state output is achieved using the minimal application of the control law. The reaching law was given by

$$s(k+1) = 0 \tag{3.23}$$

Looking at (3.23), one can notice that regardless of the size of initial condition, the sliding variable approaches to zero in the next sample instant.

Remark 3.6 The approach devised in (3.23) looks pretty simple and straightforward to implement, however, the amount of control effort required is significantly large, since the control execution task is executed in just a single sample instant.

For the perturbed system, this reaching law is modified as:

$$s(k+1) = \delta(k) \tag{3.24}$$

where $\delta(k)$ is a bounded disturbance and satisfies $|\delta(k)| \leq \delta_m$.

Remark 3.7 The size of the ultimate band depends on the size of the upper bound of the disturbance. Incidently, this band size is smaller than Gao's.

3.8.2 *Bartoszewicz Reaching Law*

This reaching law [12] differs from the one proposed in [11] in the sense that it does not require the system state to cross the sliding surface at every control step. As a result, chattering is eliminated, meaning that after the transient phase, the system state and its output remain unchanged in successive control steps. This leads to a significant reduction in control effort and enhances the performance of the quasi-sliding mode control. The reaching law conditions are stated as:

$$|s(k)| < \epsilon \implies |s(k+1)| \leq \epsilon \tag{3.25}$$

The reaching law in this was developed as:

$$s(k+1) = s_p(k+1) + \delta(k) \tag{3.26}$$

where $\delta(k)$ is a bounded disturbance and satisfies $|\delta(k)| \leq \delta_m$, $s_p(k)$ is a priori designed function and given as:

$$s_p(k) = \begin{cases} \frac{k_f - k}{k_f} s_0 & \text{for } k < k_f \\ 0 & \text{for } k \geq k_f \end{cases}$$

where s_0 is the initial condition of the sliding variable, $k_f \in \mathbb{R}_+$ and is chosen such that the $k_f < \frac{s_0}{2\delta_m}$ holds.

Remark 3.8 This type of non-switching reaching law is capable of reducing the size of ultimate band without requiring the high control effort as compared to Utkin's law. The size of the band is bounded by the upper bound of the disturbance i.e., δ_m.

3.9 Conclusion

This chapter provided a comprehensive overview of the history and evolution of discrete-time sliding mode control, starting with the foundational principles of continuous-time sliding mode control. It outlined a basic design methodology for CSMC and then traced the development of DSMC, highlighting key differences in their design criteria. A notable classification was also introduced, distinguishing between switching and non-switching behaviors of various approaches.

The next chapter will build on this foundation by examining the outcomes and limitations of these methods and introducing the proposed strategies to address these challenges.

References

1. Utkin, V.I.: Variable structure systems with sliding modes. IEEE Trans. Autom. Control **AC-22**(2), 212–222 (1977)
2. Utkin, V.I., Vadim, I.: Sliding mode control. In: Variable structure systems: from principles to implementation, p. 66 (2004)
3. Shtessel, Y., Edwards, C., Fridman, L., Levant, A.: Sliding mode control and observation, 1st edn. Springer, Birkhauser, New York (2014)
4. Fridman, L., Moreno, J.A., Bandyopadhyay, B., Kamal, S., Chalanga, A.: Continuous nested algorithms: The fifth generation of sliding mode controllers. In: Recent advances in sliding modes: From control to intelligent mechatronics, pp. 5–35. Springer (2015)
5. Utkin, V.I.: Sliding Modes in Control and Optimization. Springer-Verlag, Berlin, Germany (1992)
6. Dote, Y., Hoft, R.G.: Microprocessor based sliding mode controller for dc motor drives, presented at the Ind. Applicat. Soc. Annu. Meeting, Cincinnati, OH (1980)
7. Milosavljevic, D.: General conditions for the existence of a quasi-sliding mode on the switching hyperplane in discrete variable structure systems. Autom. Remote. Control. **46**(3), 307–314 (1985)
8. Sarpturk, S., Istefanopulos, S.Y., Kaynak, O.: On the stability of discrete-time sliding mode control systems. IEEE Trans. Automat. Contr. **32**, 930–932 (1987)
9. Furuta, K.: Sliding mode control of a discrete system. Syst. Control Lett. **14**(2), 145–152 (1990)
10. Sira-Ramirez, H.: Non-linear discrete variable structure systems in quasi-sliding mode. Int. J. Control **54**(5), 1171–1187 (1991)
11. Gao, W., Wang, Y., Homaifa, A.: Discrete-time variable structure control systems. IEEE Trans. Ind. Electron. **42**(2), 117–122 (1995)
12. Bartoszewicz, A.: Discrete-time quasi-sliding-mode control strategies. IEEE Trans. Ind. Electron. **45**(4), 633–637 (1998)
13. Utkin, V.I.: Sliding mode control in discrete-time and difference systems. Variable structure and Lyapunov control, pp. 87–107 (1994)
14. Gao, W., Hung, J.C.: Variable structure control of nonlinear systems: A new approach. IEEE Trans. Ind. Electron. **40**(1), 45–55 (1993)
15. Utkin, V.I., Drakunov, S.V.: On discrete-time sliding modes. IFAC Proc. **22**(3), 273–278 (1989)

Chapter 4
A Difference Equation with Minima Based Reaching Law for Discrete Sliding Mode Control

Abstract Here we introduce two reaching laws utilizing difference equations with minimum function for discrete-time sliding mode control, aiming to combine the advantages of Gao's and Utkin's reaching laws. These methods are developed to address the limitations of each approach–namely, Gao's law, which suffers from chattering, and Utkin's law, which may lead to excessively large control actions. The proposed reaching laws are applicable to both unperturbed and perturbed systems. For unperturbed systems, the sliding function goes to manifold in finite span of time, while in perturbed systems, it remains close to the switching manifold. The efficacy of these techniques is shown via a pendulum system example affected by matched-type bounded perturbations, with simulation results showcasing the robustness and efficiency of the proposed laws.

4.1 Introduction

This chapter introduces two reaching laws formulated on the framework of difference equations with minima for discrete-time sliding mode control. The objective here is to present an approach that strikes a balance between Gao's reaching law and Utkin's reaching law. These methods aim to address the shortcomings of both approaches, namely chattering in the Gao's method and excessively large control magnitude in the Utkin's. The proposed reaching laws for discrete-time sliding mode control are specifically designed to handle both unperturbed and perturbed systems [1]. In the case of a plant affected by disturbances, the switching function is guaranteed to converge to manifold within a finite span of time, ensuring that the system reaches the desired sliding surface efficiently. On the other hand, when perturbations are present in the system, the switching variable does not converge exactly to zero but remains in close proximity to the switching manifold, thereby maintaining the stability and robustness of the system despite the disturbances. This approach effectively balances the performance of the system in both ideal and perturbed conditions. The efficacy of the proposed methods is validated through an illustrative example involving a pendulum system affected by matched type bounded disturbances. Simulation results underscore the effectiveness of the discussed reaching law-based methodologies.

S. Kamal and P. Prasun, *A Minimum Function-Based Discrete Sliding Mode Control: Theory and Applications*, Studies in Systems, Decision and Control 621,
https://doi.org/10.1007/978-981-95-1161-7_4

This chapter is structured as follows: Sect. 4.2 offers preliminary information that will aid in understanding the key outcomes of the study. Section 4.3 presents the primary findings of this chapter. Section 4.4 validates the efficacy of the proposed reaching law based discrete sliding mode control with an example of pendulum system. Finally, Sect. 4.5 concludes this chapter.

4.2 A Brief Recollection of DSMC

This section delves into pertinent background results pertaining to the design and structure of two crucial reaching laws which consequently forms the basis for our problem formulation. The aim is to provide a comprehensive understanding of the theoretical underpinnings and practical implications of these reaching laws in the context of control system design. By exploring the foundational aspects of these reaching laws, readers can gain insights into their significance and potential applications within the broader framework of control theory. The subsequent discussions will shed light on the intricacies of these reaching laws and their role in shaping the behavior of dynamical systems.

- In the literature, as discussed in the previous chapter, there are predominantly two schools of thought to design discrete sliding mode control:
  1. Switching based control
  2. Non-switching based control
- **Switching based control** [2]**:**
  This type of control is obtained from the direct discretization of switching based control laws for continuous-time cases. In this case, the system trajectory can't get arbitrarily close to the origin due to finite switching frequency.
- **Non-switching based control** [3, 4]**:**
  The fundamental idea behind this type of development is that discrete-time control is inherently discontinuous. The equivalent control based reaching law [3] results in no reaching phase but very high amount of control action.

4.2.1 On Structural Limitation of Gao's Reaching Law

$$s(k+1) = (1 - qT)s(k) - \epsilon\, T \,\mathrm{sign}(s(k)) \tag{4.1}$$

where $q \in (0, 1)$, $\epsilon > 0$, $T > 0$, is the sampling time and $(1 - qT) > 0$ must hold which makes the choice of T quite restricted. An illustration of the evolution of sliding variable according to Gao's law is shown in Fig. 4.1. One can observe the oscillatory behaviour of the trajectory from the graph. Moreover, the following remark points out another limitation of this reaching law.

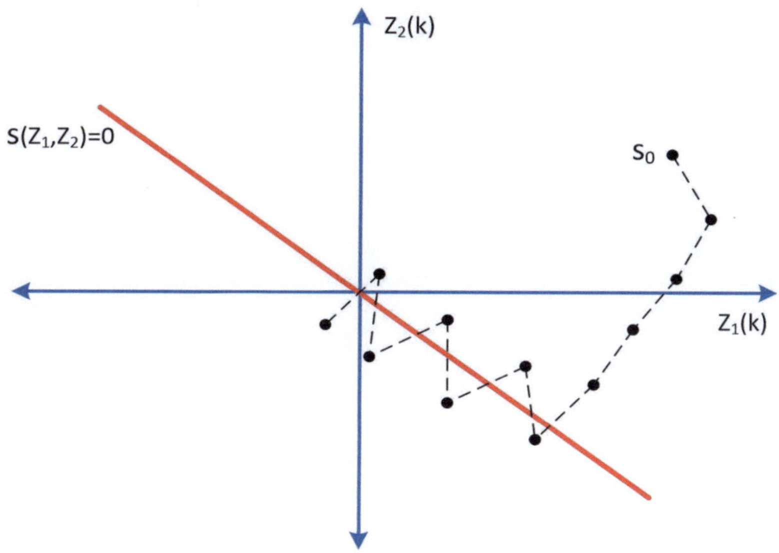

Fig. 4.1 Evolution of sliding variable according to Gao's law

Remark 4.1 Most discrete-time sliding mode control methods [2, 5, 6], such as Gao's method, result in the sliding variable oscillating around $s = 0$ rather than remaining on it. This occurs because $s = 0$ might not align with the integral multiple of the selected sampling time T, which ultimately leads to the chattering phenomenon.

4.2.2 On Structural Limitation of Utkin's Reaching Law

$$s(k+1) = 0 \tag{4.2}$$

An illustration of evolution of sliding variable according to Utkin's law is shown in Fig. 4.2. Although, this reaching law looks quite simple in its structure, however, this simple structure results in dead-beat type response. As a result, the amount of control effort required to execute this reaching law is very high.

The limitations observed in the structure of these two reaching laws motivate us to explore this area in order to craft a feasible solution to this problem.

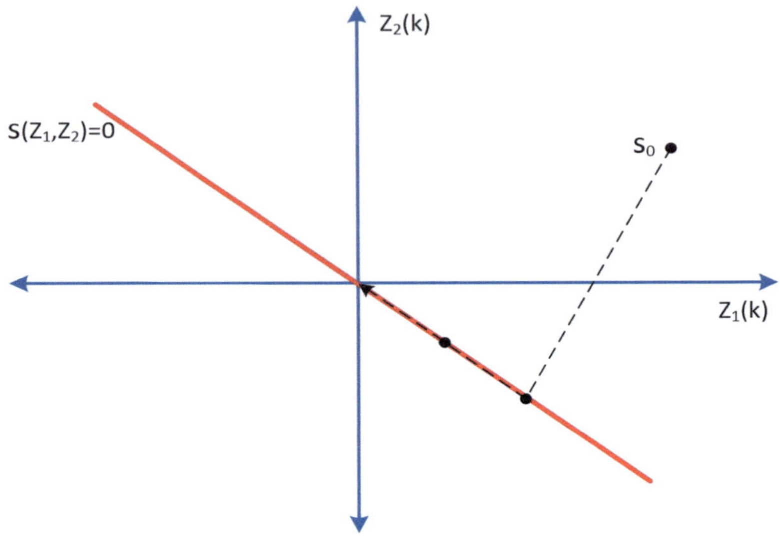

Fig. 4.2 Evolution of sliding variable according to Utkin's law

4.3 Reaching Laws for Unperturbed and Perturbed Discrete-Time Systems

In this section, we propose two reaching laws for unperturbed and perturbed discrete-time systems. We provide the definitions of sliding mode and quasi-sliding mode along with the theorems showcasing the fundamental properties of the reaching laws.

4.3.1 Reaching Law 1 for Unperturbed Discrete-Time Systems

Let us take into consideration the following reaching law for discrete-time system without disturbance:

$$s(k+1) = s(k) - \text{sign}[s(k)] \min(|s(k)|, \gamma), \tag{4.3}$$

where $\gamma \in \mathbb{R}_+$ is a positive constant chosen by the designer.

Now, consider the following discrete-time system (DTS):

$$z(k+1) = Az(k) + bu(k), \tag{4.4}$$

where $z \in D \subseteq \mathbb{R}^n$, with D being an open subset, $k \in \mathbb{Z}_{\geq 0}$, $u \in \mathbb{R}$, $A \in \mathbb{R}^{n\times n}$, and $b \in \mathbb{R}^{n\times 1}$. Sliding variable is defined as $s(k) := \wp(z(k))$. The level set $\wp^{-1}(0) := \{z \in D : \wp(z) = 0\}$ represents the switching manifold, and it is assumed to be sufficiently smooth. For our analysis, it is assumed that $\wp(z(k)) = c^\top z(k)$.

Let us take into account the following switching variable:

$$s(k) = c^\top z(k), \tag{4.5}$$

where $c \in \mathbb{R}^{n\times 1}$ and its entries are in such a way that $c^\top b \neq 0$.

Definition 4.1 There exists a sliding mode on the switching manifold $\wp^{-1}(0)$ if and only if $s(k) = 0$.

Remark 4.2 Definition 4.1 resembles the sliding mode in conventional sense, as defined for continuous-time systems. In contrast to all the definitions available in the literature for the quasi-sliding mode [2, 4], here $s(k)$ exactly reaches $\wp^{-1}(0)$ and remains there for all subsequent times.

Remark 4.3 Definition 4.1 differs from the one proposed in [2], as the reaching law introduced in our case need not the plant trajectories to cross and recross the switching manifold $\wp^{-1}(0)$, owing to its configuration. Consequently, we eliminate chattering and also steer clear of high control magnitude.

Definition 4.2 The discrete-time system (4.4) follows the reaching criteria for the sliding mode if and only if, for some $k \geq 0$, the condition below satisfies:

$$|s(k)| > \gamma \implies |s(k+1)| < |s(k)| - \nu, \tag{4.6}$$

$$|s(k)| \leq \gamma \implies s(k+1) = 0 \tag{4.7}$$

where $\nu \in \mathbb{R}_+$, is sufficiently small.

To compute $u(k)$ such that (4.3) holds, we use (4.4) and (4.5) to calculate

$$s(k+1) = c^\top Az(k) + c^\top bu(k) \tag{4.8}$$

On comparing (4.3) and (4.8), control law is obtained as $u(k) = (c^\top b)^{-1}\{-c^\top Az(k) + s(k) - \text{sign}[s(k)]\min(|s(k)|, \gamma)\}$.

Up-next, we investigate the reaching law proposed to see its attributes.

Theorem 4.1 *Consider the reaching law (4.3). If $|s(0)| > \gamma$ with $\gamma > 0$, then for some $k \geq \mathcal{K}(s_0)$, the absolute value of the sliding function of the system eventually becomes zero, where $\mathcal{K}(s_0) = \lceil \frac{|s_0|}{\gamma} \rceil$.*

Proof: The proposed reaching law (4.3) is:

$$s(k+1) = s(k) - \text{sign}[s(k)]\min(|s(k)|, \gamma).$$

Consider the case when $|s(0)| \leq \gamma$. In this scenario,

$$s(k+1) = s(k) - \text{sign}[s(k)]|s(k)| \implies s(k+1) = 0$$

in a subsequent moment, holding the conditions mentioned in Definition 4.2, with $\mathcal{K}(s_0) = 1$ time step. Moreover, for all $k \geq \mathcal{K}(s_0)$, $s(k) = 0$, demonstrating that the sliding mode exists following the Definition 4.1.

Next, consider the scenario, $|s(0)| > \gamma$. In this case,

$$s(k+1) = s(k) - \text{sign}[s(k)]\gamma.$$

We can further express this as:

$$|s(k+1)| = |s(k) - \text{sign}[s(k)]\gamma|.$$

Since $|s(k)| > \gamma$, it follows that:

$$|s(k+1)| \leq |s(k)| - \gamma.$$

Further, it can be written as

$$\begin{aligned} |s(k)| &\leq |s(k-1)| - \gamma \\ &\leq |s(k-2)| - 2\gamma \\ &\vdots \\ &\leq |s(0)| - k\gamma \end{aligned} \tag{4.9}$$

If $(|s(0)| - k\gamma) \leq \gamma$, then from (4.9), we have $|s(k)| \leq \gamma$, which implies $s(k+1) = 0$ for all $k \geq \lceil \frac{|s_0|}{\gamma} \rceil$. The expression for settling time function is thus stated as $\mathcal{K}(s_0) = \lceil \frac{|s_0|}{\gamma} \rceil$. Therefore, the modulus of $s(k)$ vanish after some finite span of time. □

Next, we consider the case where plant is affected by matched-type bounded disturbances.

4.3.2 *Reaching Law 1 for Perturbed Discrete-Time Systems*

Let us take into account the following RL for discrete-time plants affected by perturbations

$$s(k+1) = s(k) - \text{sign}[s(k)] \min(|s(k)|, \gamma) + c^\top b\delta(k) \tag{4.10}$$

where $\delta(k)$ represents the matched-type bounded perturbation (MTBP), satisfying $\underline{\delta} \leq \delta(k) \leq \bar{\delta}$, with $\underline{\delta}, \bar{\delta} \in \mathbb{R}$ being a priori known scalars. Moreover, $\delta_0 = \frac{\underline{\delta}+\bar{\delta}}{2}$ is the mean value of $\delta(k)$, and $\delta_d = \frac{\bar{\delta}-\underline{\delta}}{2}$ represents the maximum distance from the average value of perturbation. The parameter $\gamma \in \mathbb{R}_+$ is chosen such that $\gamma \geq \max\{|\underline{\delta}|, |\bar{\delta}|\}$.

To investigate the behaviour of the proposed law, let us take into consideration the DTS below which is affected by disturbances:

$$z(k+1) = Az(k) + bu(k) + b\delta(k) \tag{4.11}$$

where $z \in D \subseteq \mathbb{R}^n$, with D being an open subset, $k \in \mathbb{Z}_{\geq 0}$, $u \in \mathbb{R}$, $A \in \mathbb{R}^{n\times n}$, $b \in \mathbb{R}^{n\times 1}$, and $\delta(k)$ denotes the matched-type bounded perturbation. The sliding variable is described as $s(k) := \wp(z(k))$. The manifold $\wp^{-1}(0) := \{z \in D : \wp(z) = 0\}$ specifies the switching surface, which is assumed to be sufficiently differentiable. For the analysis, we consider $\wp(z(k)) = c^\top z(k)$, where $c \in \mathbb{R}^{n\times 1}$ is chosen such that $c^\top b \neq 0$ and $|c^\top b| \leq 1$.

Furthermore, let $\wp(z(k))$ be as defined in (4.5). For analysis purposes, let $\max\{|\underline{\delta}|, |\bar{\delta}|\} = \delta_m$, here $\delta_m \in \mathbb{R}_+$.

Definition 4.3 The quasi-sliding mode exists in the vicinity of the sliding manifold $\wp^{-1}(0)$, if and only if $|s(k)| \leq |c^\top b|\delta_m$.

Definition 4.4 A discrete-time plant mentioned by (4.11) satisfies the reaching condition for the QSM in the vicinity of sliding manifold $\wp^{-1}(0)$ if and only if, for some $k \geq 0$, the condition mentioned below satisfies:

$$|s(k)| > \gamma \implies |s(k+1)| < |s(k)| - \nu, \tag{4.12}$$

$$|s(k)| \leq \gamma \implies 0 \leq |s(k+1)| \leq |c^\top b|\delta_m \tag{4.13}$$

where ν is a positive scalar whose magnitude is sufficiently small.

To find $u(k)$ in a way that (4.10) satisfies, we make use of (4.11) and (4.5) to get

$$s(k+1) = c^\top Az(k) + c^\top bu(k) + c^\top b\delta(k) \tag{4.14}$$

On comparing (4.10) and (4.14), control law is obtained as $u(k) = (c^\top b)^{-1}$ $\{-c^\top Az(k) + s(k) - \text{sign}[s(k)]\min(|s(k)|, \gamma)\}$.

Next, we present a theorem that outlines the properties of the mentioned reaching law.

Theorem 4.2 *If in the reaching law (4.10), $|s(0)| > \gamma$, $\gamma > 0$, and $|c^\top b| \leq 1$, then for some $k \geq \mathcal{K}(s_0)$, the modulus of the switching variable of the system has ultimately bound of $|c^\top b|\delta_m$, where*

$$\mathcal{K}(s_0) = \left\lceil \frac{|s_0| - |c^\top b|\delta_m}{\gamma - |c^\top b|\delta_m} \right\rceil.$$

Proof: Rewriting the proposed reaching law in (4.10):

$$s(k+1) = s(k) - \text{sign}[s(k)] \min(|s(k)|, \gamma) + c^\top b\delta(k).$$

If we consider the case when $|s(0)| \leq \gamma$, then

$$s(k+1) = s(k) - \text{sign}[s(k)]|s(k)| + c^\top b\delta(k) \implies s(k+1) = c^\top b\delta(k)$$

in just one sampling instant. Furthermore, it follows that

$$|s(k+1)| \leq |c^\top b|\delta_m,$$

which satisfies the criterion as per Definition 4.4, with finite span as $\mathcal{K}(s_0) = 1$ time step. Hence, for every $k \geq \mathcal{K}(s_0)$, we have

$$|s(k)| \leq |c^\top b|\delta_m,$$

indicating that quasi-sliding mode exists in accordance with Definition 4.3. Now, the next scenario the system would face is when $|s(0)| > \gamma \implies s(k+1) = s(k) - \text{sign}[s(k)]\gamma + c^\top b\delta(k)$. We can further express this as

$$|s(k+1)| = |s(k) - \text{sign}[s(k)]\gamma + c^\top b\delta(k)|.$$

Since $|s(k)| > \gamma$, it follows that

$$|s(k+1)| \leq |s(k)| - \gamma + |c^\top b|\delta_m.$$

Furthermore, we can write it as

$$\begin{aligned} |s(k)| &\leq |s(k-1)| - \gamma + |c^\top b|\delta_m \\ &\leq |s(k-2)| - 2(\gamma - |c^\top b|\delta_m) \\ &\vdots \\ &\leq |s(0)| - k(\gamma - |c^\top b|\delta_m) \end{aligned} \tag{4.15}$$

If $(|s(0)| - k(\gamma - |c^\top b|\delta_m)) \leq \gamma$, then from (4.15), we have $|s(k)| \leq \gamma \implies |s(k+1)| \leq |c^\top b|\delta_m$ for all $k \geq \left\lceil \frac{|s_0| - |c^\top b|\delta_m}{\gamma - |c^\top b|\delta_m} \right\rceil$. Further, the expression for settling time function is given by $\mathcal{K}(s_0) = \left\lceil \frac{|s_0| - |c^\top b|\delta_m}{\gamma - |c^\top b|\delta_m} \right\rceil$. Therefore, we conclude that modulus of the switching variable has the steady-state bound of $|c^\top b|\delta_m$, which is achieved after the finite time calculated above. We now conclude this proof. □

Remark 4.4 We find it worth mentioning that the size of the ultimate band, $|s(k)| \leq |c^\top b|\delta_m \leq \delta_m$, could be made further smaller with appropriately selecting our design choice c, guaranteeing that $|c^\top b| \leq 1$.

4.3.3 Reaching Law 2 for Unperturbed Discrete-Time System

We now take into consideration the following reaching law for unperturbed discrete-time plant unaffected by any perturbation:

$$s(k+1) = s(k) - \gamma\, \text{sign}[s(k)] \min\left(\frac{|s(k)|}{\gamma}, |s(k)|^{\varpi}\right) \tag{4.16}$$

where $\gamma \in \mathbb{R}_+$ is a design parameter which is selected by the user, and $\varpi \in (0, 1)$.

Remark 4.5 In the reaching law (4.16), if γ made 1 and ϖ is put as a small value near to zero, then (4.16) has tendency similar to reaching law (4.3) with $\gamma = 1$.

Remark 4.6 The tuning of parameters γ and ϖ has direct effect on the convergence of the sliding variable, as shown in Fig. 4.3. The slanting of the curves nearby $s(k) = 0$ depicts slower convergence (blue and red curves), where γ and ϖ have lower values. On the other hand, keeping ϖ nearby one and large value of γ results in more rapid convergence of the switching function. In addition to this, the choice of γ and ϖ for other plants should be made keeping in mind the saturation limits of actuator of these plants.

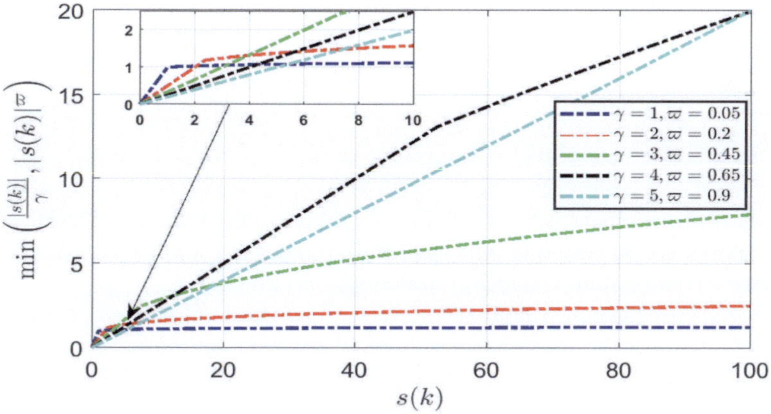

Fig. 4.3 Evolution of the min(·, ·) function with change in γ and ϖ

We now take into account the discrete-time plant as described by (4.4) and the sliding variable as given by (4.5). The sliding mode occurrence happens following Definition 4.1.

Definition 4.5 The discrete-time system (4.4) follows the reaching criterion for the sliding mode iff, for a time step $k \geq 0$, the given condition satisfies:

$$\frac{|s(k)|}{\gamma} > |s(k)|^{\varpi} \implies |s(k+1)| < |s(k)| - \nu, \tag{4.17}$$

$$\frac{|s(k)|}{\gamma} \leq |s(k)|^{\varpi} \implies s(k+1) = 0 \tag{4.18}$$

where ν is a positive scalar which is sufficiently small.

We utilise (4.4) and (4.5) to compute $u(k)$ such that (4.16) is followed,

$$s(k+1) = c^{\top} Az(k) + c^{\top} bu(k) \tag{4.19}$$

On comparing (4.16) and (4.19), required control law is obtained as, $u(k) = (c^{\top} b)^{-1} \left\{ -c^{\top} Az(k) + s(k) - \gamma \text{sign}[s(k)] \min\left(\frac{|s(k)|}{\gamma}, |s(k)|^{\varpi}\right) \right\}$. We examine the proposed reaching law and the properties it demonstrates.

Theorem 4.3 *If in the reaching law (4.16), we have* $\frac{|s(0)|}{\gamma} > |s(0)|^{\varpi}$, *with* $\gamma > 0$ *and* $\varpi \in (0, 1)$, *then for some* $k \geq \mathcal{K}(s_0)$, *the absolute value of the sliding function of the system will ultimately become zero, where* $\mathcal{K}(s_0) \leq \left\lceil \log_{[1-\gamma|s_0|^{\varpi-1}]} \frac{\gamma^{\frac{1}{1-\varpi}}}{|s_0|} \right\rceil + 1$.

Proof: We rewrite the proposed reaching law in (4.16) as follows:

$$s(k+1) = s(k) - \gamma \text{sign}[s(k)] \min\left(\frac{|s(k)|}{\gamma}, |s(k)|^{\varpi}\right).$$

If we consider the case when $\frac{|s(0)|}{\gamma} \leq |s(0)|^{\varpi}$, then we have

$$s(k+1) = s(k) - \text{sign}[s(k)]|s(k)| \implies s(k+1) = 0$$

in the next time moment, which satisfies the criterion mentioned in Definition 4.5, with $\mathcal{K}(s_0) = 1$, i.e., only one time step. Furthermore, for all $k \geq \mathcal{K}(s_0)$, $s(k) = 0$, indicating the sliding mode occurs following Definition 4.1.

Now, consider the case when $\frac{|s(0)|}{\gamma} > |s(0)|^{\varpi}$, which implies

$$s(k+1) = s(k) - \gamma \text{sign}[s(k)]|s(k)|^{\varpi}.$$

We can rewrite this as

$$|s(k+1)| = |s(k) - \gamma \text{sign}[s(k)]|s(k)|^{\varpi}|.$$

Since $|s(k)| > \gamma |s(k)|^{\varpi}$, it follows that

$$|s(k+1)| \le |s(k)| - \gamma |s(k)|^{\varpi}.$$

Further, we have the following:

$$\begin{aligned} |s(k)| &\le |s(k-1)|(1 - \gamma |s(k-1)|^{\varpi-1}) \\ &\le |s(k-2)|(1 - \gamma |s(k-2)|^{\varpi-1})(1 - \gamma |s(k-1)|^{\varpi-1}) \\ &\vdots \\ &\le |s(0)|(1 - \gamma |s(0)|^{\varpi-1}) \cdots (1 - \gamma |s(k-1)|^{\varpi-1}). \end{aligned}$$

Since $\varpi \in (0, 1)$, we have that $(1 - \gamma |s(k-1)|^{\varpi-1}) < (1 - \gamma |s(k-2)|^{\varpi-1})$, and so on. Therefore, we can write:

$$|s(k)| \le |s(0)|(1 - \gamma |s(0)|^{\varpi-1})^k. \tag{4.20}$$

If $\left||s(0)|(1 - \gamma |s(0)|^{\varpi-1})^k\right| \le \gamma |s(k)|^{\varpi}$, then from (4.20), we have $|s(k)| \le \gamma |s(k)|^{\varpi}$, which implies that $s(k+1) = 0$ for all $k \ge \left\lceil \log_{[1-\gamma |s_0|^{\varpi-1}]} \frac{\gamma^{\frac{1}{1-\varpi}}}{|s_0|} \right\rceil + 1$.

The settling time function is denoted as $\mathcal{K}(s_0) \le \left\lceil \log_{[1-\gamma |s_0|^{\varpi-1}]} \frac{\gamma^{\frac{1}{1-\varpi}}}{|s_0|} \right\rceil + 1$.

Hence, the modulus of the switching variable attains zero value post the above calculated finite span of time. We now conclude the proof. □

Next, we consider the scenario in which the plant is affected by a matched-type bounded disturbance.

4.3.4 Reaching Law 2 for Perturbed Discrete-Time Systems

We now take into consideration the reaching law mentioned below for a discrete-time plant affected by perturbation:

$$s(k+1) = s(k) - \gamma \text{sign}[s(k)] \min\left(\frac{|s(k)|}{\gamma}, |s(k)|^{\varpi}\right) + c^{\top} b \delta(k) \tag{4.21}$$

Here, $\delta(k)$ represents the matched type bounded disturbance satisfying $\underline{\delta} \le \delta(k) \le \bar{\delta}$, where $\underline{\delta}, \bar{\delta} \in \mathbb{R}$ are known constants. Additionally, $\delta_0 = \frac{\underline{\delta}+\bar{\delta}}{2}$, which represents the average of $\delta(k)$, and $\delta_d = \frac{\bar{\delta}-\underline{\delta}}{2}$, which denotes the maximum distance from the average of $\delta(k)$. The scalar constant $\gamma \in \mathbb{R}_+$ is selected in such a way that $\gamma \geq \max\{|\underline{\delta}|, |\bar{\delta}|\}$ holds.

To investigate the attributes of the proposed law, we take into account the plant given by (4.11) along with the switching function defined in (4.5).

The QSM occurrence is defined in accordance with Definition 4.3.

Definition 4.6 The DTS (4.11) holds the reaching criterion for the QSM in the vicinity of the sliding manifold iff, for some time step $k \geq 0$, the below condition is met:

$$\frac{|s(k)|}{\gamma} > |s(k)|^{\varpi} \implies |s(k+1)| < |s(k)| - \nu, \tag{4.22}$$

$$\frac{|s(k)|}{\gamma} \leq |s(k)|^{\varpi} \implies 0 \leq |s(k+1)| \leq |c^{\top}b|\delta_m \tag{4.23}$$

where ν denotes a sufficiently small positive constant.

To compute $u(k)$ such that (4.21) is satisfied, we utilize (4.11) and (4.5):

$$u(k) = (c^{\top}b)^{-1}\Big\{ - c^{\top}Az(k) + s(k) - \gamma \text{sign}[s(k)] \min\left(\frac{|s(k)|}{\gamma}, |s(k)|^{\varpi}\right)\Big\}.$$

The upcoming theorem aims to detail the attributes of the reaching law (4.21).

Theorem 4.4 *If, in the RL (4.21),* $\frac{|s(0)|}{\gamma} > |s(0)|^{\varpi}$, $\gamma > 0$, $\varpi \in (0, 1)$, *and* $|c^{\top}b| \leq 1$, *then for some* $k \geq \mathcal{K}(s_0)$, *the modulus of the switching variable of the plant has steady-state bound of* $|c^{\top}b|\delta_m$. *Here,* $\mathcal{K}(s_0) \leq \left\lceil \log_{[1-\gamma|s_0|^{\varpi-1}+|c^{\top}b|\delta_m|s_0|^{-1}]} \frac{\gamma^{\frac{1}{1-\varpi}}}{|s_0|} \right\rceil + 1$.

Proof: The proposed law is (4.21):

$$s(k+1) = s(k) - \gamma \text{sign}[s(k)] \min\left(\frac{|s(k)|}{\gamma}, |s(k)|^{\varpi}\right) + c^{\top}b\delta(k).$$

If we consider the case where $\frac{|s(0)|}{\gamma} \leq |s(0)|^{\varpi}$, then

$$s(k+1) = s(k) - \text{sign}[s(k)]|s(k)| + c^{\top}b\delta(k) \implies s(k+1) = c^{\top}b\delta(k),$$

in the next time moment. Furthermore, $|s(k+1)| \leq |c^{\top}b|\delta_m$ holds, satisfying the criterion as per Definition 4.6, with finite time as $\mathcal{K}(s_0) = 1$. Additionally, for all $k \geq \mathcal{K}(s_0)$, $|s(k)| \leq |c^{\top}b|\delta_m$, demonstrating the existence of the QSM as per Definition 4.3.

Next, consider the case where $\frac{|s(0)|}{\gamma} > |s(0)|^{\varpi}$, which implies

$$s(k+1) = s(k) - \gamma \text{sign}[s(k)]|s(k)|^{\varpi} + c^{\top} b\delta(k).$$

We can further write:

$$|s(k+1)| = \left|s(k) - \gamma \text{sign}[s(k)]|s(k)|^{\varpi} + c^{\top} b\delta(k)\right|.$$

Since $|s(k)| > |s(k)|^{\varpi}$, it follows that

$$|s(k+1)| \leq |s(k)| - \gamma |s(k)|^{\varpi} + |c^{\top} b|\delta_m.$$

This can be further expressed as:

$$\begin{aligned}
|s(k)| &\leq |s(k-1)|(1 - \gamma |s(k-1)|^{\varpi-1} + |c^{\top} b|\delta_m |s(k-1)|^{-1}) \\
&\leq |s(k-2)|(1 - \gamma |s(k-2)|^{\varpi-1} + |c^{\top} b|\delta_m |s(k-2)|^{-1}) \\
&\quad (1 - \gamma |s(k-1)|^{\varpi-1} + |c^{\top} b|\delta_m |s(k-1)|^{-1}) \\
&\quad \vdots \\
&\leq |s(0)|(1 - \gamma |s(0)|^{\varpi-1} + |c^{\top} b|\delta_m |s(0)|^{-1}) \cdots \\
&\quad (1 - \gamma |s(k-1)|^{\varpi-1} + |c^{\top} b|\delta_m |s(k-1)|^{-1})
\end{aligned} \tag{4.24}$$

Since $\varpi \in (0, 1)$, it follows that

$$(1 - \gamma |s(k-1)|^{\varpi-1} + |c^{\top} b|\delta_m |s(k-1)|^{-1}) < (1 - \gamma |s(k-2)|^{\varpi-1} + |c^{\top} b|\delta_m |s(k-2)|^{-1})$$

and so on. Thus, we have

$$|s(k)| \leq |s(0)| \left(1 - \gamma |s(0)|^{\varpi-1} + |c^{\top} b|\delta_m |s(0)|^{-1}\right)^k. \tag{4.25}$$

If

$$\left| |s(0)| \left(1 - \gamma |s(0)|^{\varpi-1} + |c^{\top} b|\delta_m |s(0)|^{-1}\right)^k \right| \leq \gamma |s(k)|^{\varpi},$$

then from (4.25), $|s(k)| \leq \gamma |s(k)|^{\varpi}$, which implies

$$|s(k+1)| \leq |c^{\top} b|\delta_m$$

for all

$$k \geq \left\lceil \log_{[1-\gamma|s_0|^{\varpi-1}+|c^{\top} b|\delta_m|s_0|^{-1}]} \frac{\gamma^{\frac{1}{1-\varpi}}}{|s_0|} \right\rceil + 1,$$

with the settling time function given by

$$\mathcal{K}(s_0) \leq \left\lceil \log_{[1-\gamma|s_0|^{\varpi-1}+|c^\top b|\delta_m|s_0|^{-1}]} \frac{\gamma^{\frac{1}{1-\varpi}}}{|s_0|} \right\rceil + 1.$$

Therefore, the modulus of the switching variable goes inside steady-state bound of $|c^\top b|\delta_m$ post above calculated settling time. We now conclude the proof. □

Remark 4.7 The primary distinction between RL1 and RL2 lies in the number of design parameters, with RL2 offering a greater degree of flexibility. Furthermore, in RL2, both arguments of the minimum function are continuous functions, resulting in a comparatively smoother evolution of the sliding variable.

4.4 A Pendulum System: An Illustrative Example

Here we develop the control law utilizing reaching methods proposed in this chapter, for a pendulum system affected by disturbance. The plant under consideration is as follows:

$$\begin{aligned} z_1(k+1) &= z_1(k) + z_2(k) \\ z_2(k+1) &= z_2(k) - \frac{mgl}{2J}\sin(z_1(k)) - \frac{B}{J}z_2(k) + \frac{1}{J}u(k) + \delta(k) \end{aligned} \tag{4.26}$$

where $z_1(k)$ represents the angular position of the pendulum from the equilibrium state, angular velocity is denoted as $z_2(k)$, m is pendulum's mass, gravitational acceleration is represented as g, length of the pendulum as l, J is the inertia of the pendulum arm, B is the friction coefficient, and $u(k)$ is the control signal. The bounded disturbance $\delta(k)$ is taken as $0.1\ \sin(0.3k)$, which introduces oscillatory behavior into the system dynamics.

To control the system, we design a control law based on the principles of RL1 and RL2. These laws are chosen to ensure the system reaches a sliding mode, even in the presence of the bounded perturbation. In particular, the RL control law aims to adjust the control input $u(k)$ so that the sliding variable $s(k)$ converges to zero, ensuring the system stabilizes around the desired equilibrium point.

The perturbation $\delta(k)$ is assumed to be bounded, satisfying the condition $\underline{\delta} \leq \delta(k) \leq \bar{\delta}$, with $\underline{\delta} = -0.1$ and $\bar{\delta} = 0.1$. The proposed control law is designed to handle such perturbations and ensure that the sliding mode is maintained despite the oscillations in the perturbation signal. The control law is validated through simulation results, which demonstrate the effectiveness of the proposed RL-based approach in driving the system to the desired state while compensating for the bounded perturbations.

For simulation purposes, the plant variables have the following values: $m = 1.1\,\text{kg}$, $l = 1\,\text{m}$, $g = 9.81\ \frac{\text{m}}{\text{s}^2}$, and $B = 0.18\ \frac{\text{kg}\cdot\text{m}}{\text{s}^2}$. The initial conditions and design parameters are given by $z_1(0) = 5$, $z_2(0) = 2$, $\gamma = 6$, $c_1 = 0.3$, $c_2 = 1$, and $\alpha = 0.5$.

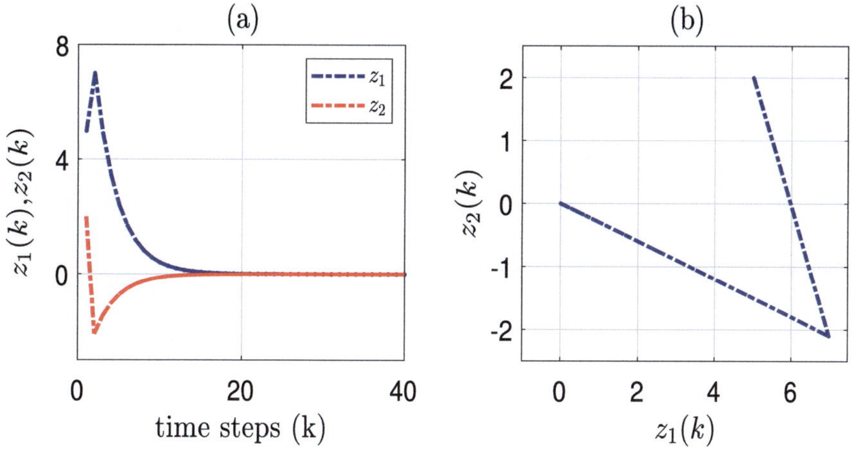

Fig. 4.4 States and phase portrait for RL1, in the case of unperturbed DTS

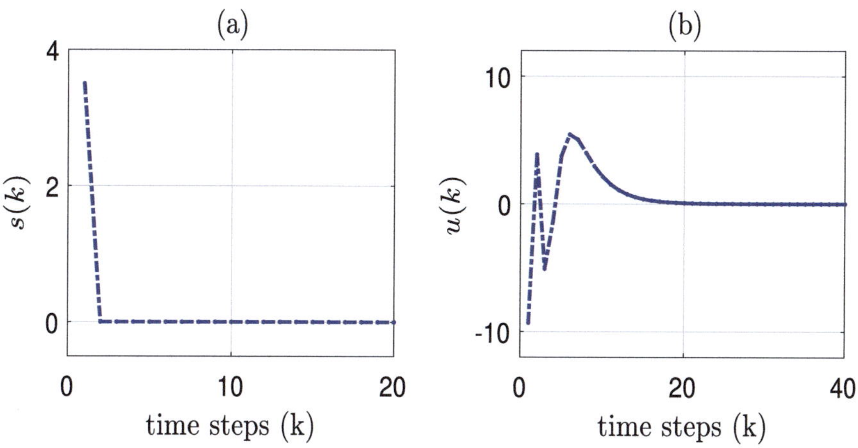

Fig. 4.5 Sliding variable and control input for RL1, in the case of unperturbed DTS

Figures 4.4, 4.5, 4.6 and 4.7 correspond to reaching law 1. The time response of the plant trajectories for unperturbed scenario is shown in Fig. 4.4, while the switching function and control signal are presented in Fig. 4.5. We note that both the switching variable and the control input reach zero within a finite span of time.

The time response of the plant trajectories when system is affected by disturbances, switching function and control signal are shown in Figs. 4.6 and 4.7. Because of the occurrence of the disturbance, the state trajectories and switching function lie inside a bound throughout the process, leading to a non-zero control input.

Figures 4.8, 4.9, 4.10 and 4.11 correspond to RL2. Similar to RL1, the states, sliding variable, and control input converge to zero, as seen in Figs. 4.8 and 4.9. For

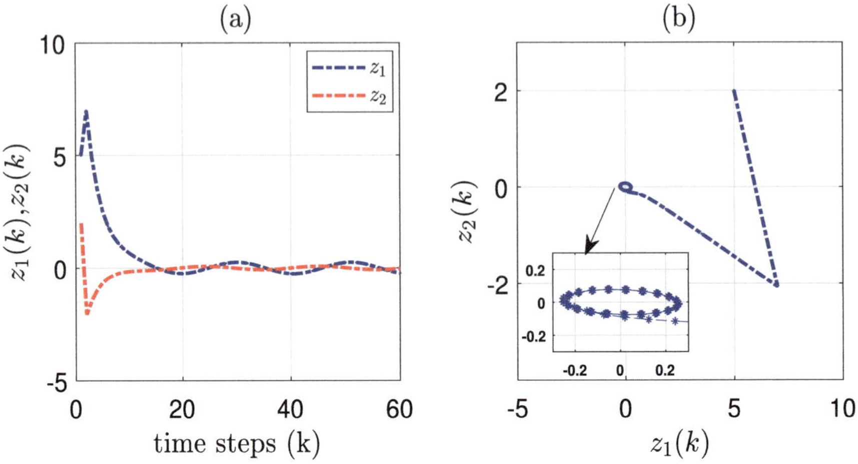

Fig. 4.6 States and phase portrait for RL1, in the case of perturbed DTS

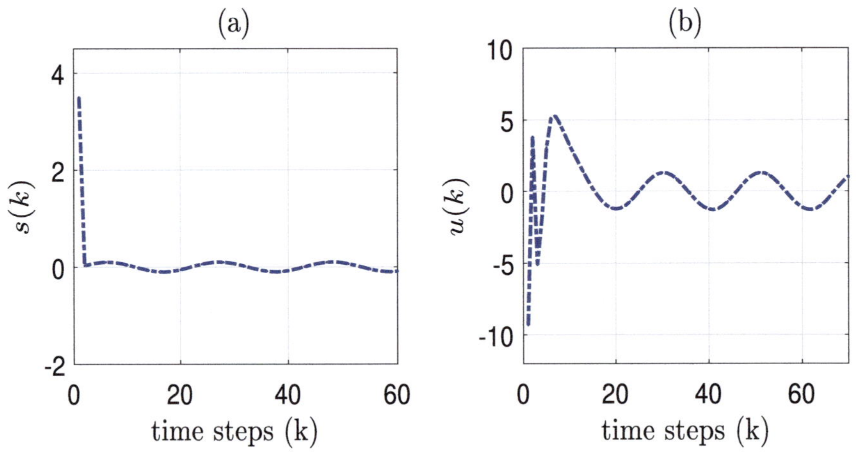

Fig. 4.7 Sliding variable and control input for RL1, in the case of perturbed DTS

the perturbed case, depicted in Figs. 4.10 and 4.11, the states and sliding variable remain within an invariant set, while the control input remains non-zero.

4.5 Conclusion

In this chapter, we have conducted a thorough investigation of reaching laws for discrete variable structure systems, utilizing a difference equation involving minimum function. This novel approach is designed to overcome the drawbacks of both Gao's

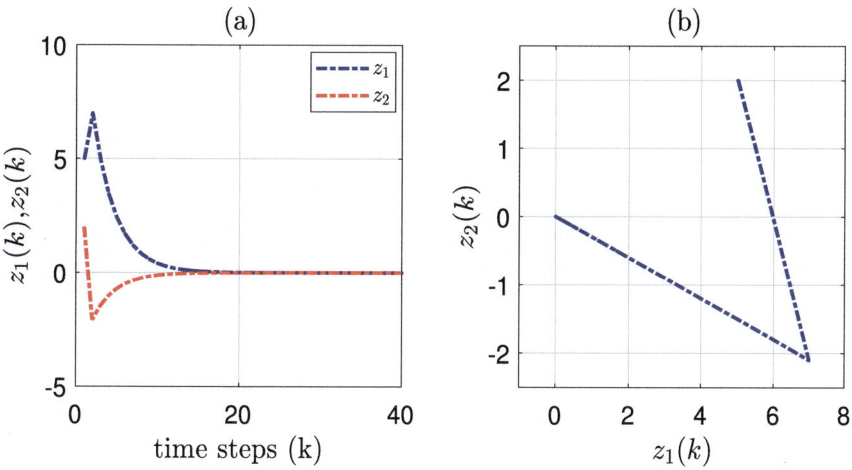

Fig. 4.8 States and phase portrait for RL2, in the case of unperturbed DTS

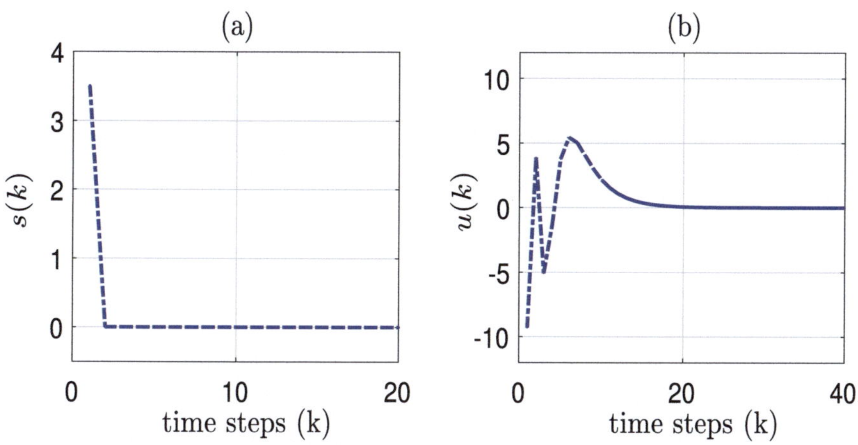

Fig. 4.9 Sliding variable and control input for RL2, in the case of unperturbed DTS

reaching law and Utkin's equivalent control-based technique. The reaching laws discussed here addressed two crucial challenges: the complete elimination of chattering and the prevention of excessive control actions. To depict the effectiveness of these laws, simulation results are presented using a pendulum system as an example. The simulations cover both unperturbed and perturbed scenarios, offering a comprehensive assessment of the proposed methods. The simulation outcomes provide strong evidence of the efficacy and robustness of the introduced reaching laws.

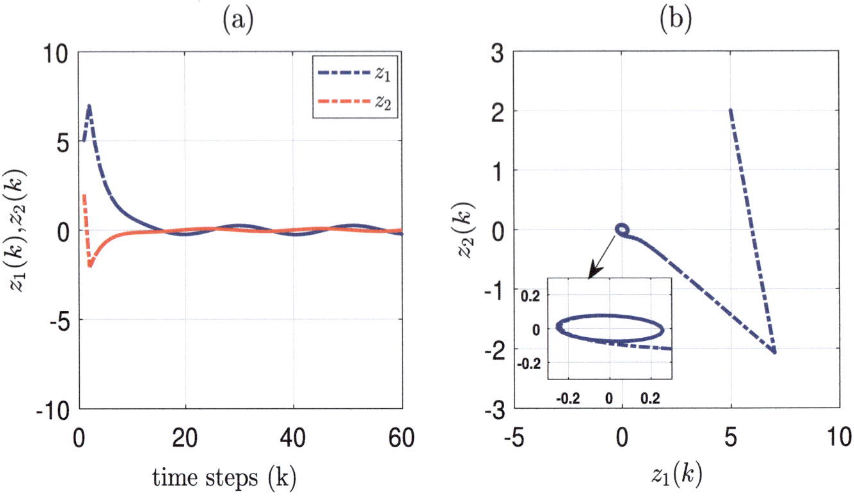

Fig. 4.10 States and phase portrait for RL2 in the presence of perturbed DTS

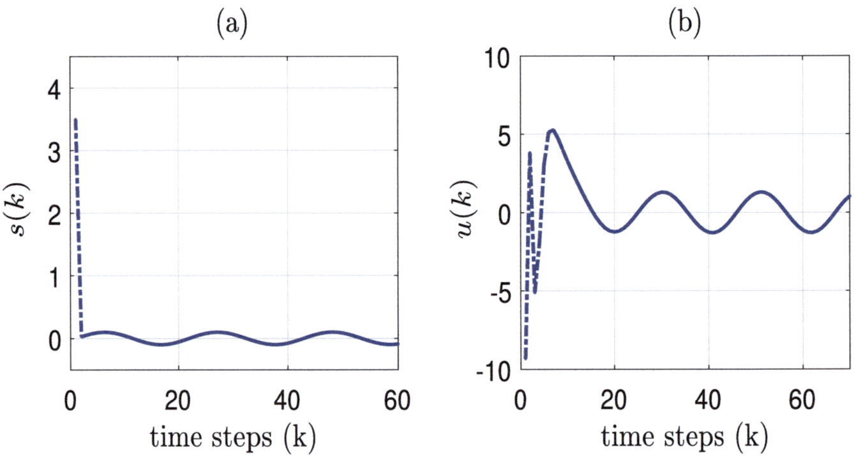

Fig. 4.11 Sliding variable and control input for RL2 in the presence of perturbed DTS

In the next chapter, the scope of reaching laws utilizing the difference equation with minimum function will be extended by incorporating a variable gain through a proportional term, utilizing a rate-regulatory function. This extension aims to tackle the issue of slow convergence rates encountered in the current approach.

References

1. Prasun, P., Kamal, S., Bartoszewicz, A., Ghosh, S.: A difference equation with minima-based reaching law for discrete variable structure systems, IEEE Trans. Circ. Syst. II: Expr. Briefs **71**(3), 1236–1240 (2023)
2. Gao, W., Wang, Y., Homaifa, A.: Discrete-time variable structure control systems. IEEE Trans. Ind. Electron. **42**(2), 117–122 (1995)
3. Utkin, V.I., Drakunov, S.V.: On discrete-time sliding modes. IFAC Proc. **22**(3), 273–278 (1989)
4. Bartoszewicz, A.: Discrete-time quasi-sliding-mode control strategies. IEEE Trans. Ind. Electron. **45**(4), 633–637 (1998)
5. Bartoszewicz, A., Lesniewski, P.: New switching and nonswitching type reaching laws for SMC of discrete time systems. IEEE Trans. Control Syst. Tech. **24**(2), 670–677 (2016)
6. Bartoszewicz, A., Adamiak, K.: Discrete-time sliding-mode control with a desired switching variable generator. IEEE Trans. Autom. Control **65**(4), 1807–1814 (2020)

Chapter 5
Reaching Laws Based on Difference Equations with Minima and Rate-Regulatory Function

Abstract This chapter presents reaching laws utilizing difference equation with minimum function for discrete-time sliding mode control. The proposed approach combines the characteristics of an altered Gao's methodology when away from the sliding manifold and Utkin's method when nearby the manifold. By eliminating chattering and ensuring precise adherence of the system states to the sliding hyperplane, the method effectively controls the rate of change of the switching function through a tunable gain. Consequently, the derived control signal is less aggressive for large initial conditions compared to Gao's method. The proposed discrete-time sliding mode control is applicable to both unperturbed and perturbed systems. For unperturbed systems, the sliding variable reaches zero in finite time, whereas for perturbed systems, it remains confined within a neighborhood of the sliding hyperplane. The effectiveness of the proposed reaching law is validated through simulations, including both a numerical example and a practical example involving a pendulum system. Additionally, a detailed comparative analysis with existing methods and the approach discussed in the previous chapter is provided, offering a comprehensive evaluation.

5.1 Introduction

This chapter presents modified reaching laws making use of difference equation with minimum function for discrete-time SMC. The proposed method integrates the features of altered Gao's method when the system operates farther from the sliding manifold and Utkin's method when the system is near the origin. This approach effectively eliminates chattering and ensures precise adherence of the plant trajectories to the switching manifold. Furthermore, it regulates the variation of the switching function through a tunable gain, resulting in a control signal that is less aggressive for large initial conditions compared to Gao's reaching law [1, 2].

The described reaching law is applicable to system without disturbances as well as system affected by disturbances. For unperturbed systems, the switching function converges to sliding manifold in finite span of time, while for systems affected by perturbation, it remains confined within a neighborhood of the sliding hyperplane. The efficacy of the discussed method is confirmed via both numerical and practical

S. Kamal and P. Prasun, *A Minimum Function-Based Discrete Sliding Mode Control: Theory and Applications*, Studies in Systems, Decision and Control 621,
https://doi.org/10.1007/978-981-95-1161-7_5

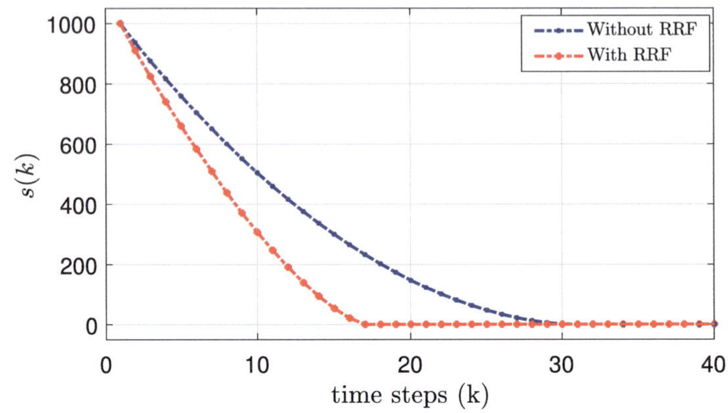

Fig. 5.1 A comparative depiction of the sliding variable's evolution

examples, including the simulation of a pendulum setup. Additionally, a detailed comparative analysis with existing methods and those introduced in previous chapters provides a comprehensive evaluation.

The remainder of the chapter is organized as follows: Sect. 5.2 presents the motivation for incorporating rate-regulatory functions (RRF). Section 5.3 outlines the main results of this chapter. In Sect. 5.4, the effectiveness of the proposed scheme is depicted with the help of two specimens: a numerical system and a practical pendulum system. Finally, Sect. 5.5 provides the concluding remarks for this chapter.

5.2 Rate-Regulatory Function

The following results will be helpful in order to understand the concept of variable gain based rate of change of sliding variable.

With the help of rate-regulatory function [2], we intend to change the evolution rate of the sliding variable by incorporating the former with the proportional term in the reaching law. An illustration of this is shown in Fig. 5.1. We can see that reaching law without RRF tend to converge much slowly as compared with that of the RRF based reaching law.

The corresponding curve for such type of functions is shown in Fig. 5.2. One can notice that, the curve in the graph has value close to 1 for large values of the sliding variable, which results in the less rate of change of the sliding variable to reduce the amount of control effort required. Similarly, for smaller values of the sliding variable, the curve approaches 0, resulting in a significantly faster rate of change of the sliding variable.

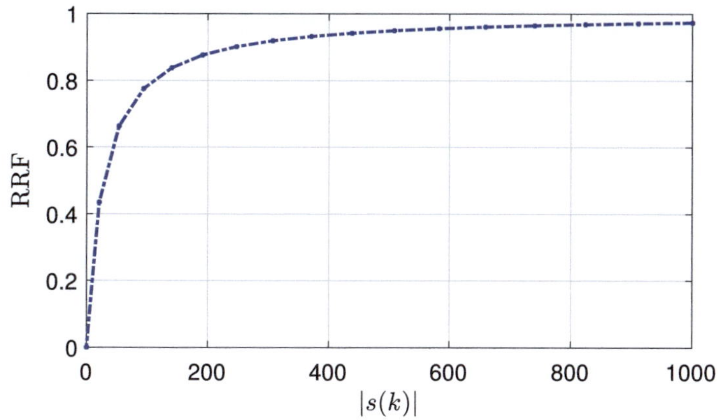

Fig. 5.2 Variation of rate-regulatory function with respect to $s(k)$

A comparative analysis is performed between the reaching law proposed in the previous chapter and the RRF-based reaching law. Through simulation, it is demonstrated that the RRF-based reaching law exhibits superior convergence properties compared to the algorithm introduced in the previous chapter. For the purpose of simulation, all common parameters were kept consistent, with $\gamma = 2$. The initial condition was also identical, i.e., $s(0) = 1000$.

From Fig. 5.1, it can be observed that the method discussed here achieves good convergence results as we compare these from the previous chapter. This improvement is attributed to the inclusion of the rate-regulatory function (RRF). In contrast, the algorithm from the previous chapter lacks an RRF and instead employs a fixed factor of one, leading to slower motion towards the maniflod. The attributes of the RRF as a map with argument as s are illustrated in Fig. 5.2.

Notably, the presence of the RRF allows the proportional term's coefficient to be dynamically adjusted within the full range of $[0, 1)$. For large initial conditions, the RRF approaches 1, effectively limiting the variation rate of the switching function. Conversely, for values near the sliding manifold, the RRF decreases towards 0, enabling a faster response. This dynamic adjustment imparts significantly improved convergence properties to the RRF-based reaching law.

In the following section, RRF functions are utilized to design reaching laws making use of difference equation with minimum function, aiming to achieve improved transient performance.

5.3 Rate-Regulatory Function Based Reaching Laws for DSMC

Here we introduce reaching laws for discrete-time plants without disturbance as well as when affected by the disturbance, utilizing difference equations with minimum function, incorporating the rate-regulatory function.

5.3.1 *Reaching Law 1 for Unperturbed Discrete-Time Systems*

Let us take into account the following reaching law for discrete-time plant without perturbation

$$s(k+1) = \Xi(s)s(k) - \operatorname{sgn}[s(k)]\min(\Xi(s)|s(k)|, \gamma) \tag{5.1}$$

Here, $\Xi(s) = 1 - \lambda(s)$, where the function $\lambda : \mathbb{R} \to (0, 1]$ is expressed as $\lambda(s) = \frac{\varrho}{\varrho + |s(k)|}$, and $\varrho, \gamma \in \mathbb{R}_+$ are design parameters selected by the user.

Remark 5.1 The function λ is designed so that the value $\Xi(s)$ becomes high for high values of $|s(k)|$ and reduces for lower magnitude of $|s(k)|$, as shown in Fig. 5.3. It results in a rapid variation rate of $s(k)$ for low $|s(k)|$ and a slower variation rate for bigger $|s(k)|$.

Remark 5.2 The map $\Xi(s)$ is selected based on the work in [3], in which authors refined the proportional term from [4], thereby ensuring the plant's variables remain on the feasible set.

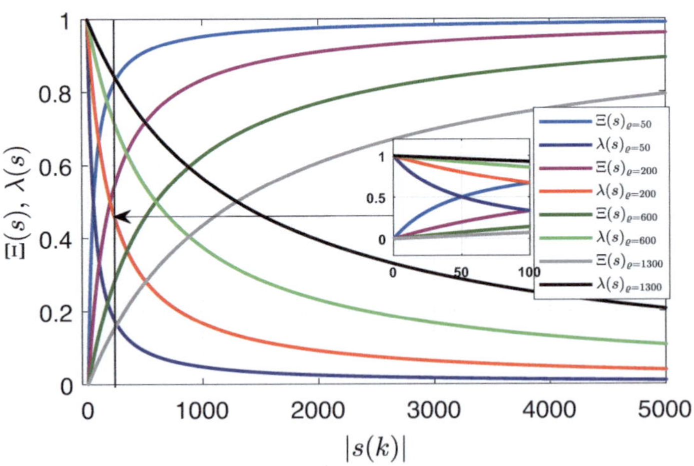

Fig. 5.3 $\Xi(s)$ and $\lambda(s)$ for different values of ϱ

Remark 5.3 The behavior of the mapping $\Xi(s)$ for different values of ϱ is such that it is flatter towards 0 for smaller $|s(k)|$ and flatter towards 1 for larger $|s(k)|$, as shown in Fig. 5.3. Choosing smaller values of ϱ results in $\Xi(s)$ being closer to 1, which slows down the system response. In contrast, selecting larger values of ϱ keeps $\Xi(s)$ significantly less than 1, causing it to decrease rapidly towards zero and resulting in a faster system response.

According to the proposed reaching law (5.1), the rate of change of the sliding variable can be expressed as

$$\begin{aligned}\Delta s(k) =& |s(k+1) - s(k)| \\ =& |\Xi(s)s(k) - \text{sgn}[s(k)] \min(\Xi(s)|s(k)|, \gamma) - s(k)| \\ =& | - \lambda(s)s(k) - \text{sgn}[s(k)] \min(\Xi(s)|s(k)|, \gamma)|. \end{aligned} \tag{5.2}$$

In the case where the plant trajectories are farther from the switching manifold, we have $\min(\Xi(s)|s(k)|, \gamma) = \gamma$. Consequently, it follows that

$$\begin{aligned}\Delta s(k) =& | - \lambda(s)s(k) - \text{sgn}[s(k)]\gamma| \\ =& \left| -\frac{\varrho}{\varrho + |s(k)|} s(k) - \text{sgn}[s(k)]\gamma \right| \\ \leq& \varrho + \gamma. \end{aligned} \tag{5.3}$$

Therefore, the variation rate of the switching function is consistently constrained by the user-defined constants.

Remark 5.4 In the described law (5.1), it is evident that for any k, the switching function $s(k)$ forever lie inside a region fixed by the user-defined gains, specifically $\varrho + \gamma$, irrespective of the magnitude of the initial conditions. This is a great advantage of the described reaching law, as the boundary is decided by the designer.

Now, we take into consideration the following plant

$$z(k+1) = Az(k) + bu(k) \tag{5.4}$$

where $z \in D \subseteq \mathbb{R}^n$, D is an open subset, $k \in \mathbb{Z}_{\geq 0}$, $u \in \mathbb{R}$, $A \in \mathbb{R}^{n \times n}$, and $b \in \mathbb{R}^{n \times 1}$. We describe the sliding variable as $s(k) := \wp(z(k))$. The region $\wp^{-1}(0) := \{z \in D : \wp(z) = 0\}$ represents the switching manifold, which is sufficiently differentiable. For the purpose of this analysis, we further consider that $\wp(z(k)) = c^\top z(k)$. Hence, the switching function is given by

$$s(k) = c^\top z(k) \tag{5.5}$$

where $c \in \mathbb{R}^{n \times 1}$ is selected so that $c^\top b \neq 0$.

Definition 5.1 We assert that sliding mode exists on the sliding manifold $\wp^{-1}(0)$ if $s(k) = 0$.

Remark 5.5 Definition 5.1 shares a similarity with the traditional sliding mode, which was given for continuous-time systems. In contrast to the existing definitions of the quasi-sliding mode [4, 5], this approach ensures that the switching function precisely converges to the sliding manifold $\wp^{-1}(0)$.

Remark 5.6 Furthermore, Definition 5.1 deviates from the standard definition, as proposed in [4], because the approach introduced in these results does not need the plant trajectories to cross and recross the switching manifold $\wp^{-1}(0)$. This characteristic is present because of the configuration of the discussed reaching law. Consequently, chattering is completely removed, which was a common issue in [4], and unnecessary control magnitude is not needed. Additionally, the potential for stimulating the unmodeled high-frequency dynamics in the plant is mitigated.

Definition 5.2 A discrete-time system (5.4) follows the reaching criterion for sliding mode iff, for any $k \geq 0$, the given conditions hold:

$$\Xi(s)|s(k)| > \gamma \implies |s(k+1)| < |s(k)| - \nu, \tag{5.6}$$

$$\Xi(s)|s(k)| \leq \gamma \implies s(k+1) = 0, \tag{5.7}$$

where ν is a sufficiently small positive constant.

To determine $u(k)$ such that (5.1) holds, we use (5.4) and (5.5) to calculate

$$s(k+1) = c^\top Az(k) + c^\top bu(k) \tag{5.8}$$

On comparing (5.1) and (5.8), the control law is obtained as

$$u(k) = (c^\top b)^{-1}\{-c^\top Az(k) + \Xi(s)s(k) - \text{sgn}[s(k)]\min(\Xi(s)|s(k)|, \gamma)\} \tag{5.9}$$

Further, we examine the developed reaching law and the attributes it demonstrates.

Theorem 5.1 *Consider the reaching law* (5.1). *If* $\Xi(s_0)|s(0)| > \gamma$ *with* $\gamma > 0$, *then for some* $k \geq \mathcal{K}(s_0)$, *it holds that* $s(k) = 0$, *where* $\mathcal{K}(s_0) = \left\lceil \frac{|s_0|}{\gamma} \right\rceil$.

Proof: We rewrite the proposed RL in (5.1) as:

$$s(k+1) = \Xi(s)s(k) - \text{sgn}[s(k)]\min(\Xi(s)|s(k)|, \gamma).$$

Consider the case when $\Xi(s_0)|s(0)| \leq \gamma$. In this case, we have:

$$s(k+1) = \Xi(s)s(k) - \text{sgn}[s(k)]\Xi(s)|s(k)| \implies s(k+1) = 0$$

after the next sample moment, thus following the criterion according to Definition 5.2 with $\mathcal{K}(s_0) = 1$ time step. Furthermore, for all $k \geq \mathcal{K}(s_0)$, $s(k) = 0$, indicating that sliding mode exists in accordance with Definition 5.1.

Now, consider the case when $\Xi(s_0)|s(0)| > \gamma$. In this case, we have:

$$s(k+1) = \Xi(s)s(k) - \text{sgn}[s(k)]\gamma.$$

This implies:

$$|s(k+1)| = |\Xi(s)s(k) - \text{sgn}[s(k)]\gamma|.$$

Since $\Xi(s)|s(k)| > \gamma$ and $|\Xi(s)| \leq 1$, it follows that:

$$|s(k+1)| \leq |s(k)| - \gamma.$$

Further,

$$\begin{aligned} |s(k)| &\leq |s(k-1)| - \gamma \\ &\leq |s(k-2)| - 2\gamma \\ &\vdots \\ &\leq |s(0)| - k\gamma \end{aligned} \tag{5.10}$$

If $(|s(0)| - k\gamma) \leq \gamma$ and since $|\Xi(s)| \leq 1$, then from (5.10), we have $|s(k)| \leq \gamma \implies \Xi(s)|s(k)| \leq \gamma$, which implies that $s(k+1) = 0$ for all $k \geq \left\lceil \frac{|s_0|}{\gamma} \right\rceil$. The settling time is given by the function $\mathcal{K}(s_0) = \left\lceil \frac{|s_0|}{\gamma} \right\rceil$. Therefore, the modulus of the switching variable gets to 0 post a finite number of time steps, completing the proof. □

Upcoming, we consider the scenario in which the considered plant is affected by matched-type bounded perturbation.

5.3.2 Reaching Law 1 for Discrete-Time Systems with Perturbations

We take into account the following reaching law for discrete-time systems with perturbation:

$$s(k+1) = \Xi(s)s(k) - \text{sgn}[s(k)]\min(\Xi(s)|s(k)|, \gamma) + c^\top b\delta(k) \tag{5.11}$$

where $\delta(k)$ represents the disturbance of bounded and matched-type satisfying $\underline{\delta} \leq \delta(k) \leq \bar{\delta}$, with $\underline{\delta}, \bar{\delta} \in \mathbb{R}$ being known bounds. Moreover, let $\delta_0 = \frac{\underline{\delta}+\bar{\delta}}{2}$ denote the average of $\delta(k)$, and $\delta_d = \frac{\bar{\delta}-\underline{\delta}}{2}$ represent the maximum distance from the average value of $\delta(k)$. The scalar constant $\gamma \in \mathbb{R}_+$ is selected so that $\gamma \geq \max\{|\underline{\delta}|, |\bar{\delta}|\}$.

To investigate the attributes of the reaching law (5.11), we take into account the following perturbed discrete-time plant:

$$z(k+1) = Az(k) + bu(k) + b\delta(k) \tag{5.12}$$

where $z \in D \subseteq \mathbb{R}^n$, with D being an open subset, $k \in \mathbb{Z}_{\geq 0}$, $u \in \mathbb{R}$, $A \in \mathbb{R}^{n\times n}$, $b \in \mathbb{R}^{n\times 1}$, and $\delta(k)$ representing the matched-type bounded perturbation.

We define the switching manifold as the region $\wp^{-1}(0)$, and the sliding variable is given by (5.5), where $c \in \mathbb{R}^{n\times 1}$ with its components chosen such that $c^\top b \neq 0$ and $|c^\top b| \leq 1$. For the purpose of analysis, let $\max\{|\underline{\delta}|, |\bar{\delta}|\} = \delta_m$, where $\delta_m \in \mathbb{R}_+$.

Definition 5.3 We affirm that quasi-sliding mode exists in the vicinity of the sliding manifold $\wp^{-1}(0)$ if $|s(k)| \leq |c^\top b|\delta_m$.

Definition 5.4 The DTS (5.12) follows the reaching criterion of the QSM in the vicinity of sliding manifold iff, for any $k \geq 0$, the given condition satisfies:

$$\Xi(s)|s(k)| > \gamma \implies |s(k+1)| < |s(k)| - \nu, \tag{5.13}$$

$$\Xi(s)|s(k)| \leq \gamma \implies 0 \leq |s(k+1)| \leq |c^\top b|\delta_m \tag{5.14}$$

where ν is a sufficiently small positive constant.

Remark 5.7 The reaching law proposed in (5.1) does not account for any perturbation term, leading to the existence of an ideal sliding mode. In contrast, for the perturbed plant (5.27), the RL (5.11) describes the aftereffects of disturbance by incorporating an additional term, $c^\top b\delta(k)$.

To determine $u(k)$ such that (5.11) is satisfied, we use (5.12) and (5.5) to calculate it as

$$s(k+1) = c^\top Az(k) + c^\top bu(k) + c^\top b\delta(k) \tag{5.15}$$

On comparing (5.11) and (5.15), the control law is obtained as

$$u(k) = (c^\top b)^{-1}\{-c^\top Az(k) + \Xi(s)s(k) - \operatorname{sgn}[s(k)]\min(\Xi(s)|s(k)|, \gamma)\} \tag{5.16}$$

In the following theorem, we outline the properties of the previously discussed reaching law.

Theorem 5.2 *If, in the reaching law* (5.11), $\Xi(s_0)|s(0)| > \gamma$ *with* $\gamma > 0$ *and* $|c^\top b| \leq 1$, *then for some* $k \geq \mathcal{K}(s_0)$, *we have* $|s(k)| \leq |c^\top b|\delta_m$, *where* $\mathcal{K}(s_0) = \left\lceil \frac{|s_0| - |c^\top b|\delta_m}{\gamma - |c^\top b|\delta_m} \right\rceil$.

Proof: Rewriting the reaching law proposed in (5.11):

$$s(k+1) = \Xi(s)s(k) - \mathrm{sgn}[s(k)]\min(\Xi(s)|s(k)|, \gamma) + c^\top b\delta(k).$$

If we consider the case where $\Xi(s_0)|s(0)| \leq \gamma$, then

$$s(k+1) = \Xi(s)s(k) - \mathrm{sgn}[s(k)]\Xi(s)|s(k)| + c^\top b\delta(k) \implies s(k+1) = c^\top b\delta(k),$$

which occurs in the next time moment. Additionally, we get

$$|s(k+1)| \leq |c^\top b|\delta_m,$$

following the criterion in accordance Definition 5.4, with $\mathcal{K}(s_0) = 1$ time step. For every $k \geq \mathcal{K}(s_0)$, we have $|s(k)| \leq |c^\top b|\delta_m$, indicating the quasi-sliding mode exists according to Definition 5.3.

Now, moving to the scenario where $\Xi(s_0)|s(0)| > \gamma$, we have

$$s(k+1) = \Xi(s)s(k) - \mathrm{sgn}[s(k)]\gamma + c^\top b\delta(k).$$

We can further write

$$|s(k+1)| = |\Xi(s)s(k) - \mathrm{sgn}[s(k)]\gamma + c^\top b\delta(k)|.$$

Since $\Xi(s)|s(k)| > \gamma$, $|\Xi(s)| \leq 1$, and $|c^\top b| \leq 1$, it follows that

$$|s(k+1)| \leq |s(k)| - \gamma + |c^\top b|\delta_m.$$

Further, we can write that

$$\begin{aligned}
|s(k)| &\leq |s(k-1)| - \gamma + |c^\top b|\delta_m \\
&\leq |s(k-2)| - 2(\gamma - |c^\top b|\delta_m) \\
&\;\;\vdots \\
&\leq |s(0)| - k(\gamma - |c^\top b|\delta_m)
\end{aligned} \tag{5.17}$$

If $(|s(0)| - k(\gamma - |c^\top b|\delta_m)) \leq \gamma$ and since $|\Xi(s)| \leq 1$, then from (5.17), we have

$$|s(k)| \leq \gamma \implies \Xi(s)|s(k)| \leq \gamma \implies |s(k+1)| \leq |c^\top b|\delta_m,$$

for all $k \geq \lceil \frac{|s_0| - |c^\top b|\delta_m}{\gamma - |c^\top b|\delta_m} \rceil$, with the settling time function expressed by

$$\mathcal{K}(s_0) = \lceil \frac{|s_0| - |c^\top b|\delta_m}{\gamma - |c^\top b|\delta_m} \rceil.$$

Therefore the modulus of the switching variable has steady-state bound of $|c^{\top}b|\delta_m$ after finite span of time steps. We now conclude the proof. □

5.3.3 Reaching Law 2 for Unperturbed Discrete-Time Systems

We now take into consideration the following reaching law for an discrete-time system without perturbation

$$s(k+1) = \Xi(s)s(k) - \gamma \operatorname{sgn}[s(k)] \min\left(\frac{\Xi(s)|s(k)|}{\gamma}, |s(k)|^{\varpi}\right) \tag{5.18}$$

$$\Xi(s) = 1 - \lambda(s(k)),$$

where the mapping $\lambda : \mathbb{R} \to (0, 1]$ is expressed by

$$\lambda(s(k)) = \frac{\varrho}{\varrho + |s(k)|},$$

and $\varrho, \gamma \in \mathbb{R}_{+}$ are positive gains selected by the user, while $\varpi \in (0, 1)$. The function $\Xi(s)$ serves as a rate-regulatory function, which modulates the variation rate of the switching function $s(k)$.

We now take into account the following discrete-time plant

$$z(k+1) = Az(k) + bu(k) \tag{5.19}$$

where $z \in D \subseteq \mathbb{R}^n$, with D as an open subset, $k \in \mathbb{Z}_{\geq 0}$, $u \in \mathbb{R}$, $A \in \mathbb{R}^{n\times n}$, and $b \in \mathbb{R}^{n\times 1}$ with the assumption that the pair (A, b) is controllable. Furthermore, the sliding variable is defined as

$$s(k) := \wp(z(k)).$$

Definition 5.5 The level set $\wp^{-1}(0)$ is defined as the set of all points z for which the value of the sliding variable is 0, i.e.,

$$\wp^{-1}(0) := \{z \in D : \wp(z) = 0\}.$$

The level set $\wp^{-1}(0)$ represents the switching manifold, which is assumed to be continuously differentiable. For our analysis, we take into account a linear sliding function. Specifically, we consider the following sliding variable as:

$$s(k) = c^{\top}z(k), \tag{5.20}$$

where $c \in \mathbb{R}^{n\times 1}$, and its entries are chosen such that $c^{\top}b \neq 0$.

Definition 5.6 Sliding mode is said to exist on the sliding manifold $\wp^{-1}(0)$ if, for any $k \geq \mathcal{K}(s_0)$, $s(k) = 0$, where $\mathcal{K}(s_0)$ denotes the settling time expression.

Definition 5.7 The reaching criterion for the sliding mode in the DTS (5.19) is satisfied iff, for any $k \geq 0$, the given condition satisfies:

$$\frac{\Xi(s)|s(k)|}{\gamma} > |s(k)|^{\varpi} \implies |s(k+1)| < |s(k)| - \nu, \tag{5.21}$$

$$\frac{\Xi(s)|s(k)|}{\gamma} \leq |s(k)|^{\varpi} \implies s(k+1) = 0 \tag{5.22}$$

where ν is a sufficiently small number with positive sign.

To compute $u(k)$ such that (5.18) is satisfied, we use (5.19) and (5.20) to determine

$$s(k+1) = c^{\top} Az(k) + c^{\top} bu(k) \tag{5.23}$$

On comparing (5.18) and (5.23), control law is obtained as

$$\begin{aligned} u(k) =& (c^{\top} b)^{-1} \Big\{ - c^{\top} Az(k) + \varpi(s)s(k) - \gamma \text{sgn}[s(k)] \\ & \min \left(\frac{\Xi(s)|s(k)|}{\gamma}, |s(k)|^{\varpi} \right) \Big\} \end{aligned} \tag{5.24}$$

Upnext, we investigate the described reaching law and the attributes it demonstrates.

Theorem 5.3 *If in the reaching law* (5.18), $\frac{\Xi(s_0)|s_0|}{\gamma} > |s_0|^{\varpi}$, *with* $\gamma > 0$, $\varpi \in (0, 1)$, *then for some* $k \geq \mathcal{K}(s_0)$, $s(k) = 0$, *where* $\mathcal{K}(s_0) \leq \left\lceil \log_{[1-\gamma|s_0|^{\varpi-1}]} \frac{\gamma^{\frac{1}{1-\varpi}}}{|s_0|} \right\rceil + 1$.

Proof: Rewriting the reaching law proposed in (5.18) as:

$$s(k+1) = \Xi(s)s(k) - \gamma \text{ sgn}[s(k)] \min \left(\frac{\Xi(s)|s(k)|}{\gamma}, |s(k)|^{\varpi} \right).$$

If we consider the case when $\frac{\Xi(s_0)|s_0|}{\gamma} \leq |s_0|^{\varpi}$, then

$$s(k+1) = \Xi(s)s(k) - \text{sgn}[s(k)]\Xi(s)|s(k)| \implies s(k+1) = 0$$

after the next time moment, following the criterion as per Definition 5.7, with $\mathcal{K}(s_0) = 1$ time step. moreover, for all $k \geq \mathcal{K}(s_0)$, $s(k) = 0$, showcasing that sliding mode exits according to Definition 5.6.

Now, moving on to the scenario when $\frac{\Xi(s_0)|s_0|}{\gamma} > |s_0|^{\varpi}$, we have

$$s(k+1) = \Xi(s)s(k) - \gamma \text{ sgn}[s(k)]|s(k)|^{\varpi}.$$

We can further write

$$|s(k+1)| = |\Xi(s)s(k) - \gamma \operatorname{sgn}[s(k)]|s(k)|^{\varpi}|.$$

Since $\Xi(s)|s(k)| > \gamma|s(k)|^{\varpi}$ and $|\Xi(s)| \leq 1$, it follows that

$$|s(k+1)| \leq |s(k)| - \gamma|s(k)|^{\varpi}.$$

Furthermore,

$$\begin{aligned}
|s(k)| &\leq |s(k-1)|(1 - \gamma|s(k-1)|^{\varpi-1}) \\
&\leq |s(k-2)|(1 - \gamma|s(k-2)|^{\varpi-1})(1 \\
&- \gamma|s(k-1)|^{\varpi-1}) \\
&\vdots \\
&\leq |s_0|(1 - \gamma|s_0|^{\varpi-1}) \cdots (1 - \gamma|s(k-1)|^{\varpi-1})
\end{aligned}$$

Since, $\varpi \in (0, 1)$, $(1 - \gamma|s(k-1)|^{\varpi-1}) < (1 - \gamma|s(k-2)|^{\varpi-1})$ and so on. Then, it follows that

$$|s(k)| \leq |s_0|(1 - \gamma|s_0|^{\varpi-1})^k \tag{5.25}$$

If

$$\left|\left(|s_0|(1 - \gamma|s_0|^{\varpi-1})^k\right)\right| \leq \gamma|s_0|^{\varpi},$$

and since $|\varpi(s)| \leq 1$, then from (5.25),

$$|s(k)| \leq \gamma|s(k)|^{\varpi} \implies \Xi(s)|s(k)| \leq \gamma|s(k)|^{\varpi} \implies s(k+1) = 0$$

for all $k \geq \left\lceil \log_{[1-\gamma|s_0|^{\varpi-1}]} \frac{\gamma^{\frac{1}{1-\varpi}}}{|s_0|} \right\rceil + 1$, with the settling time function expressed as

$$\mathcal{K}(s_0) \leq \left\lceil \log_{[1-\gamma|s_0|^{\varpi-1}]} \frac{\gamma^{\frac{1}{1-\varpi}}}{|s_0|} \right\rceil + 1.$$

Therefore, the modulus of the sliding function gets to zero post finite span of time. □

Upcoming, we conduct the investigation on plant with perturbation.

5.3.4 Reaching Law 2 for Discrete-Time Systems with Perturbations

Let us take into consideration the following reaching law for discrete-time plants affected by the perturbation

$$s(k+1) = \Xi(s)s(k) - \gamma \operatorname{sgn}[s(k)] \min\left(\frac{\Xi(s)|s(k)|}{\gamma}, |s(k)|^{\varpi}\right) + c^{\top} b\delta(k) \tag{5.26}$$

where $\delta(k)$ represents the matched type bounded perturbation, satisfying $\underline{\delta} \leq \delta(k) \leq \bar{\delta}$, with $\underline{\delta}, \bar{\delta} \in \mathbb{R}$ being known bounds. Moreover, $\delta_0 = \frac{\underline{\delta}+\bar{\delta}}{2}$ is the average value of $\delta(k)$, and $\delta_d = \frac{\bar{\delta}-\underline{\delta}}{2}$ is the maximum distance from the average of $\delta(k)$. The scalar constant $\gamma \in \mathbb{R}_{+}$ is a user selected gain chosen so that $\gamma \geq \max\{|\underline{\delta}|, |\bar{\delta}|\}$.

To analyze the behavior of the reaching law (5.26), we consider the following system:

$$z(k+1) = Az(k) + bu(k) + b\delta(k) \tag{5.27}$$

where $z \in D \subseteq \mathbb{R}^n$ with D being an open subset, $k \in \mathbb{Z}_{\geq 0}$, $u \in \mathbb{R}$, $A \in \mathbb{R}^{n\times n}$, $b \in \mathbb{R}^{n\times 1}$, and $\delta(k)$ denotes the matched type bounded perturbation.

In a similar manner, the sliding manifold is defined by $\wp^{-1}(0)$, and the switching function is given by (5.20), where $c \in \mathbb{R}^{n\times 1}$ is selected such that $c^{\top} b \neq 0$ and an additional condition $|c^{\top} b| \leq 1$ is imposed (rationale behind will be discussed later). We proceed with the analysis taking into consideration $\max\{|\underline{\delta}|, |\bar{\delta}|\} = \delta_m$, where $\delta_m \in \mathbb{R}_{+}$.

Definition 5.8 We define the existence of the QSM in the vicinity of the level set $\wp^{-1}(0)$ if, for any $k \geq \mathcal{K}(s_0)$, the condition $|s(k)| \leq |c^{\top} b|\delta_m$ holds, where $\mathcal{K}(s_0)$ represents the settling time function.

Definition 5.9 The reaching condition for the QSM in the neighbourhood of $\wp^{-1}(0)$ for the system (5.27) is satisfied if and only if, for some $k \geq 0$, the following condition holds:

$$\frac{\Xi(s)|s(k)|}{\gamma} > |s(k)|^{\varpi} \implies |s(k+1)| < |s(k)| - \nu, \tag{5.28}$$

$$\frac{\Xi(s)|s(k)|}{\gamma} \leq |s(k)|^{\varpi} \implies 0 \leq |s(k+1)| \leq |c^{\top} b|\delta_m \tag{5.29}$$

where ν is a sufficiently small positive real number.

To determine $u(k)$ such that (5.26) is satisfied, we make use of (5.27) and (5.20) to calculate

$$s(k+1) = c^{\top}Az(k) + c^{\top}bu(k) + c^{\top}b\delta(k) \tag{5.30}$$

On comparing (5.26) and (5.30), control law is obtained as

$$\begin{aligned} u(k) =& (c^{\top}b)^{-1}\Big\{ -c^{\top}Az(k) + \Xi(s)s(k) - \gamma \operatorname{sgn}[s(k)] \\ & \min\left(\frac{\Xi(s)|s(k)|}{\gamma}, |s(k)|^{\varpi}\right)\Big\} \end{aligned} \tag{5.31}$$

The upcoming theorem describes the attributes of the previously mentioned method for the plants with perturbation.

Theorem 5.4 *Consider the reaching law* (5.26) *with the condition* $|c^{\top}b| \leq 1$. *If in* (5.26), $\frac{\Xi(s_0)|s_0|}{\gamma} > |s_0|^{\varpi}$, *where* $\gamma > 0$ *and* $\varpi \in (0,1)$, *then for some* $k \geq \mathcal{K}(s_0)$, *we have* $|s(k)| \leq |c^{\top}b|\delta_m$, *where* $\mathcal{K}(s_0)$ *is bounded by the expression*

$$\mathcal{K}(s_0) \leq \left\lceil \log_{1-\gamma|s_0|^{\varpi-1}+|c^{\top}b|\delta_m|s_0|^{-1}} \frac{\gamma^{\frac{1}{1-\varpi}}}{|s_0|} \right\rceil + 1.$$

Proof: Rewriting the reaching law proposed in (5.26), we have:

$$s(k+1) = \Xi(s)s(k) - \gamma \operatorname{sgn}[s(k)] \min\left(\frac{\Xi(s)|s(k)|}{\gamma}, |s(k)|^{\varpi}\right) + c^{\top}b\delta(k).$$

If we consider the case where $\frac{\Xi(s_0)|s_0|}{\gamma} \leq |s_0|^{\varpi}$, then we have:

$$s(k+1) = \Xi(s)s(k) - \operatorname{sgn}[s(k)]\Xi(s)|s(k)| + c^{\top}b\delta(k) \implies s(k+1) = c^{\top}b\delta(k),$$

which occurs in the next time moment. Therefore, we obtain

$$|s(k+1)| \leq |c^{\top}b|\delta_m,$$

following the criterion stated in Definition 5.9, with $\mathcal{K}(s_0) = 1$ time step. Moreover, for all $k \geq \mathcal{K}(s_0)$, we have $|s(k)| \leq |c^{\top}b|\delta_m$, illustrating that quasi-sliding mode exists in accordance with Definition 5.8.

Next, we consider the case when $\frac{\Xi(s_0)|s_0|}{\gamma} > |s_0|^{\varpi}$, which implies:

$$s(k+1) = \Xi(s)s(k) - \gamma \operatorname{sgn}[s(k)]|s(k)|^{\varpi} + c^{\top}b\delta(k).$$

We can express this as:

$$|s(k+1)| = |\Xi(s)s(k) - \gamma \operatorname{sgn}[s(k)]|s(k)|^{\varpi} + c^{\top}b\delta(k)|.$$

Since $\Xi(s)|s(k)| > \gamma|s(k)|^{\varpi}$ and $|\Xi(s)| \leq 1$, it follows that:

$$|s(k+1)| \le |s(k)| - \gamma|s(k)|^{\varpi} + |c^{\top}b|\delta_m.$$

Further, we can write that

$$\begin{aligned}
|s(k)| &\le |s(k-1)|(1-\gamma|s(k-1)|^{\varpi-1} + |c^{\top}b|\delta_m|s(k-1)|^{-1})\\
&\le |s(k-2)|(1-\gamma|s(k-2)|^{\varpi-1} + |c^{\top}b|\delta_m|s(k-2)|^{-1})\\
&\quad(1-\gamma|s(k-1)|^{\varpi-1} + |c^{\top}b|\delta_m|s(k-1)|^{-1})\\
&\quad\vdots\\
&\le |s_0|(1-\gamma|s_0|^{\varpi-1} + |c^{\top}b|\delta_m|s_0|^{-1})\cdots\\
&\quad(1-\gamma|s(k-1)|^{\varpi-1} + |c^{\top}b|\delta_m|s(k-1)|^{-1})
\end{aligned}$$

Since $\varpi \in (0, 1)$, we have

$$(1-\gamma|s(k-1)|^{\varpi-1} + |c^{\top}b|\delta_m|s(k-1)|^{-1}) < (1-\gamma|s(k-2)|^{\varpi-1} + |c^{\top}b|\delta_m|s(k-2)|^{-1}),$$

and so on. Therefore, it follows that

$$|s(k)| \le |s_0|(1-\gamma|s_0|^{\varpi-1} + |c^{\top}b|\delta_m|s_0|^{-1})^k \tag{5.32}$$

If

$$\left||s_0|\left(1-\gamma|s_0|^{\varpi-1} + |c^{\top}b|\delta_m|s_0|^{-1}\right)^k\right| \le \gamma|s_0|^{\varpi},$$

and since $|\varpi(s)| \le 1$, then from (5.32), we have

$$|s(k)| \le \gamma|s(k)|^{\varpi} \implies \varpi(s)|s(k)| \le \gamma|s(k)|^{\varpi} \implies |s(k+1)| \le |c^{\top}b|\delta_m,$$

for all $k \ge \left\lceil \log_{[1-\gamma|s_0|^{\varpi-1}+|c^{\top}b|\delta_m|s_0|^{-1}]} \frac{\gamma^{\frac{1}{1-\varpi}}}{|s_0|} \right\rceil + 1$, with the settling time function denoted by

$$\mathcal{K}(s_0) \le \left\lceil \log_{[1-\gamma|s_0|^{\varpi-1}+|c^{\top}b|\delta_m|s_0|^{-1}]} \frac{\gamma^{\frac{1}{1-\varpi}}}{|s_0|} \right\rceil + 1.$$

Thus, the modulus of the switching variable has steady-state bound of $|c^{\top}b|\delta_m$ after a finite number of steps. □

Remark 5.8 Again, we observe that the size of the steady-state band, given by $|s(k)| \le |c^{\top}b|\delta_m \le \delta_m$, can be considerably decreased by appropriately selecting the vector c to ensure that $|c^{\top}b| \le 1$.

In the majority of the available discrete variable structure control literature, authors have considered disturbances with zero average value. In those scenarios, $\delta_m = \delta_d$, where δ_d represents the maximum distance of the disturbance from its average. The size of the ultimate band for the work in [4] is given by $|s(k)| \le 2\delta_d + \epsilon$, for [5] as $|s(k)| \leq \delta_d$, and for [6] as $|s(k)| \leq \delta_d + \epsilon$. In this context, the discussed method results in a smaller ultimate band width, demonstrating its advantages over these previous approaches.

5.4 Simulation Results and Discussion

In this section, we present a numerical example and a practical case study to illustrate the effectiveness of the proposed reaching laws. Additionally, we perform a comparative analysis to highlight the superiority of the proposed designs. We demonstrate that, for unperturbed discrete-time systems, the switching variable arrives on the sliding manifold in finite span of time, while for systems affected by perturbations, the switching variable converges to an invariant region and remains in that region for subsequent steps of time.

5.4.1 A Numerical Example

Consider the dynamical system given by

$$\begin{aligned} z_1(k+1) &= z_1(k) + z_2(k) \\ z_2(k+1) &= u(k) + \delta(k) \end{aligned} \tag{5.33}$$

where

$$z = \begin{bmatrix} z_1 \\ z_2 \end{bmatrix} \in D \subset \mathbb{R}^2,$$

$u(k)$ is the control input, and $\delta(k)$ is the matched type bounded perturbation. It is considered here as $\sin(0.5k)$. The difference equation with the minima-based reaching law (RL1) is designed according to (5.1) and (5.11). The study is performed for five different values of ϱ, specifically 30, 70, 130, 200, and 450. The plant's initial states are set as $z_1(0) = 1000$ and $z_2(0) = 300$. The value of design gains are selected as $c_1 = 0.4$, $c_2 = 1$, and $\gamma = 8$.

The state trajectories of the nominal discrete-time plant are presented in Fig. 5.4. From the figure, it is evident that for $\varrho = 30$, the system states gradually approach zero over time. In contrast, when $\varrho = 450$, the states converge to zero at a significantly faster rate. This demonstrates the effect of varying the parameter ϱ on the convergence speed of the system.

The switching variable and the control signal for the unperturbed plant are depicted in Fig. 5.5. A key point that we highlight is, as ϱ increases, both the switching variable as well as control signal vary more rapidly. Although, for low magnitude of ϱ, the variation rate is slower, which results in a slower convergence of the sliding variable and a more moderate control input. These observations reflect how the

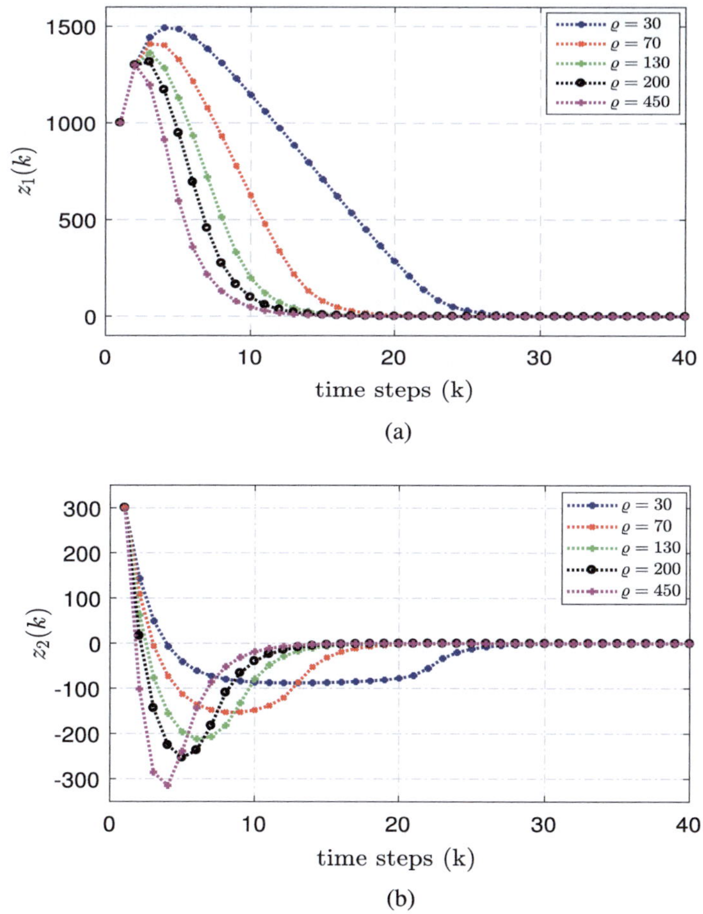

Fig. 5.4 State trjectories of the nominal discrete-time system for various values of ϱ

design parameter ϱ influences the system's behavior (Figs. 5.6 and 5.7). Figures 5.4 and 5.5 present the plant trajectories for the system with perturbation.Since plant is subjected to a time varying disturbance that follows a sine wave, the plant trajectories settle within an invariant region ultimately. Despite the perturbation, the system exhibits similar convergence properties to the unperturbed system. This highlights the robustness of the proposed approach in the presence of disturbances.

Finally, the phase portrait of the system is shown in Fig. 5.8. In the nominal scenario, the plant trajectories reduces to 0, as expected. However, for the plant with perturbation, the trajectories remain within an invariant region, illustrating the system's stability despite the disturbances. It is clear from the phase portrait that the discussed reaching law effectively forces the switching variable to the manifold in

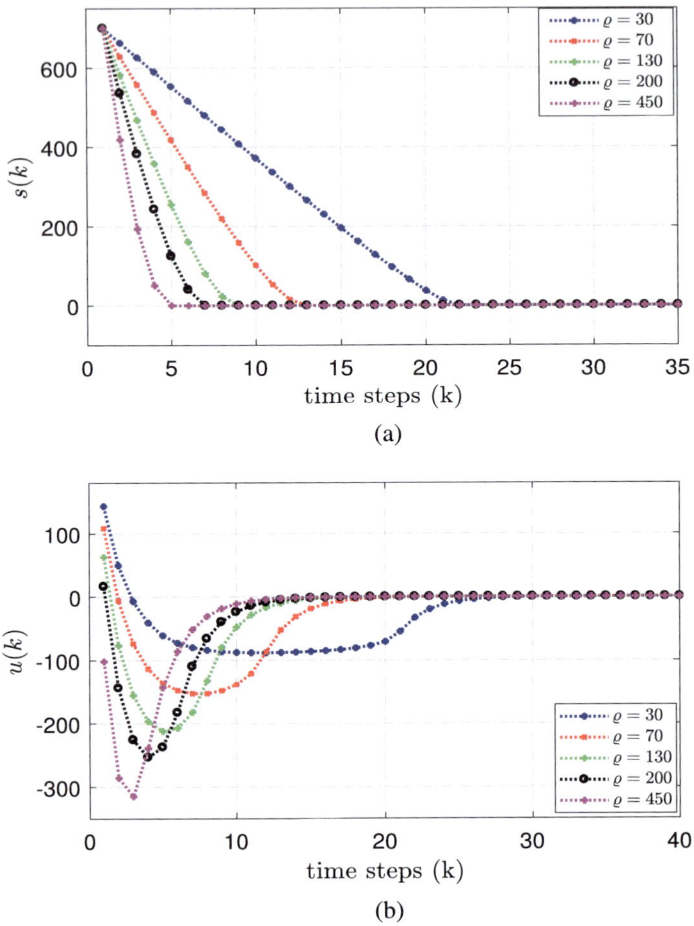

Fig. 5.5 **a** Switching variable, **b** control input, of the nominal discrete-time system for different values of ϱ

the nominal plant and to an invariant region in the perturbed DTS within a finite-time span.

Remark 5.9 The selection of an appropriate value for ϱ depends on finding a balance where the plant reaction is not excessively slow when $|s(k)|$ is less, and the control signal at the start is not much demanding. This choice is influenced by the constraints placed on the closed-loop signals.

Remark 5.10 The proposed reaching law addresses the issues of chattering and excessive control signal magnitude when $|s(k)|$ is big, as seen in [4], while also achieving $|s(k)| = 0$ for nominal discrete-time plants. Although the non-switching type reaching law in [3] resolves the chattering problem by ensuring that $s(k)$ does

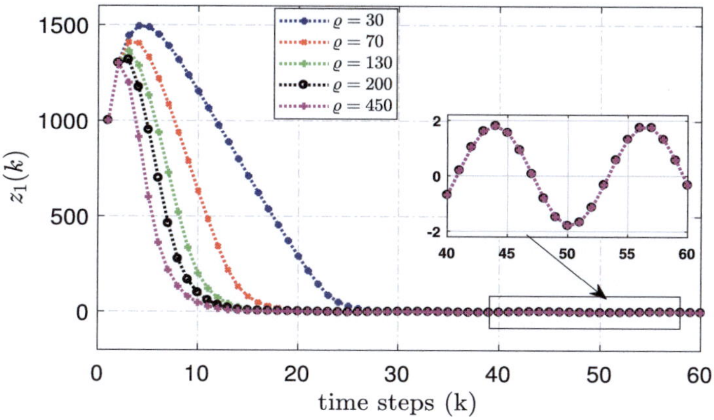

(a)

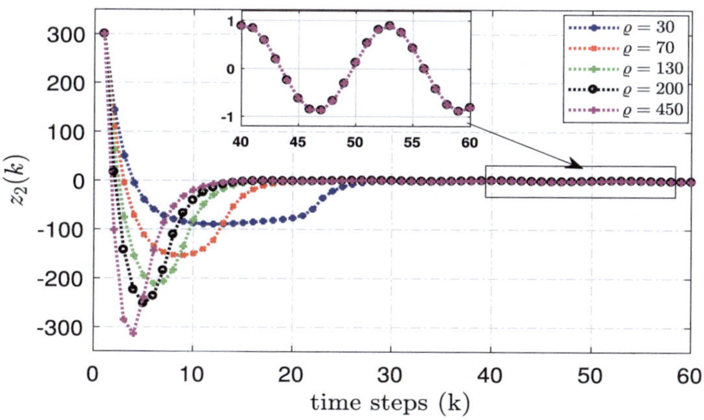

(b)

Fig. 5.6 Trajectories of the perturbed discrete-time system for various values of ϱ

not repeatedly cross the switching manifold at each control step, the discussed law offers a dead-beat type response near the sliding hyperplane, driving $s(k)$ exactly to zero.

5.4.2 Pendulum System: A Comparative Study

We now examine the example of a pendulum affected by matched-type bounded disturbances. The plant is described by:

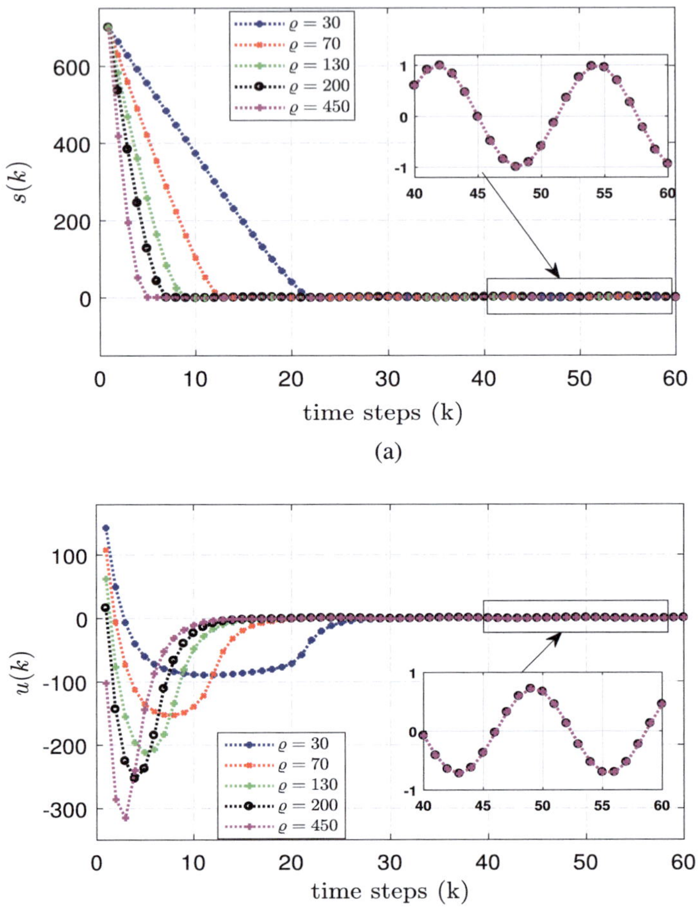

(a)

(b)

Fig. 5.7 **a** Switching variable, **b** Control input, of the perturbed discrete-time system for various values of ϱ

$$
\begin{aligned}
z_1(k+1) =& z_1(k) + z_2(k) \\
z_2(k+1) =& z_2(k) - \frac{mgl}{2J}\sin(z_1(k)) - \frac{B}{J}z_2(k) + \frac{1}{J}u(k) + \frac{1}{J}\delta(k)
\end{aligned} \tag{5.34}
$$

here, $z_1(k)$ represent the pendulum angle from the mean position, and $z_2(k)$ the angular velocity. The system parameters include m as the mass of the pendulum, g as the acceleration due to gravity, l as the length of the pendulum, J as the inertia of the pendulum arm, B as the friction coefficient, and $u(k)$ as the control signal. The bounded disturbance $\delta(k)$ is modeled as 0.05 sin(0.3k). The control signal is constructed utilizing the discussed reaching law (RL2) as described in equation (5.18). For this study via simulation, the following plant parameters are used:

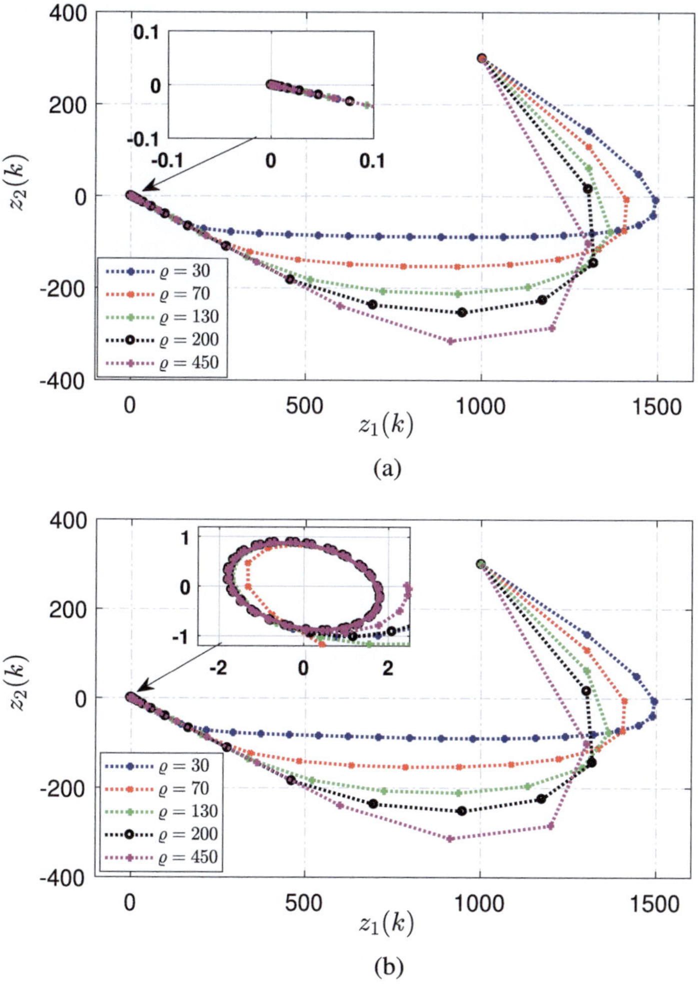

Fig. 5.8 Phase portrait: **a** Unperturbed, **b** Perturbed, discrete-time system for different values of ϱ

$m = 1.1$ kg, $l = 1$ m, $g = 9.81\ \frac{\text{m}}{\text{s}^2}$, and $B = 0.18\ \frac{\text{kg}\cdot\text{m}}{\text{s}^2}$. The initial states of the plant are set as $z_1(0) = 30$ and $z_2(0) = 15$.

To assess the effectiveness of the proposed reaching law, we conduct a comparative study against several established methods, including those proposed in [4] (denoted as R_1), [5] (R_2), and [6] (R_3), as well as the method introduced in the previous chapter (without RRF). The notations used to represent the parameters in each of these methods are selected directly from the corresponding works. The design parameters for the methods are chosen as follows: for [4], $q = 0.15$ and $\epsilon = 0.8$; for [5], $k^* = 5$; for [6], $s_0 = 25$ and $\epsilon = 0.12$; and for the proposed method (without RRF), $\gamma = 1.8$

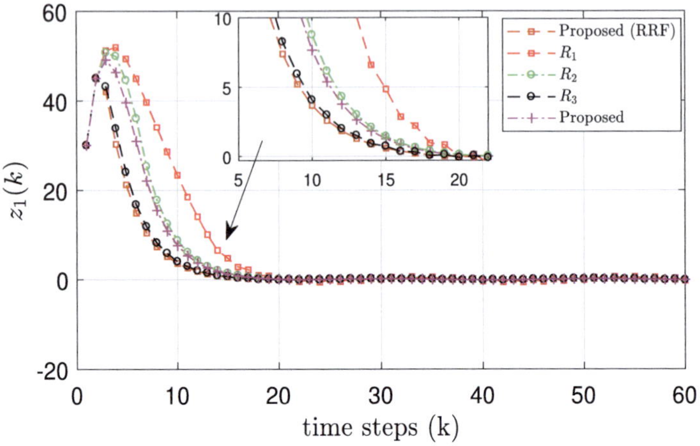

Fig. 5.9 State $z_1(k)$ of the perturbed pendulum system

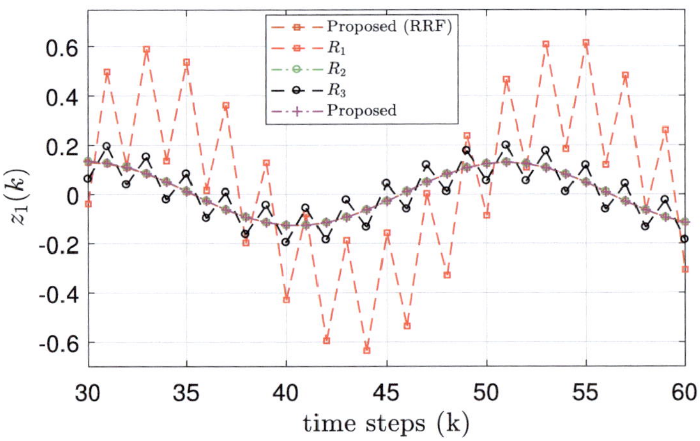

Fig. 5.10 State $z_1(k)$ during the sliding phase

and $\varpi = 0.4$. These parameters are selected based on the conditions specified in the respective references.

For the proposed reaching law with RRF, the design variables are selected as $\varrho = 10$, $\gamma = 1.8$, and $\varpi = 0.4$, with the sliding hyperplane parameters selected as $c_1 = 0.3$ and $c_2 = 0.6$.

The plant's state trajectories for different reaching laws are depicted in Figs. 5.9, 5.10, 5.11, and 5.12. In Fig. 5.9, which shows the angular position of the pendulum, it can be observed that the discussed method exhibits superior convergence attributes, having least overshoot among all the techniques. On the other hand, the system states driven by [4] and [6] display high-frequency oscillatory behavior, a consequence of their switching-based approaches, as seen in Fig. 5.10. Although [6] utilizes the

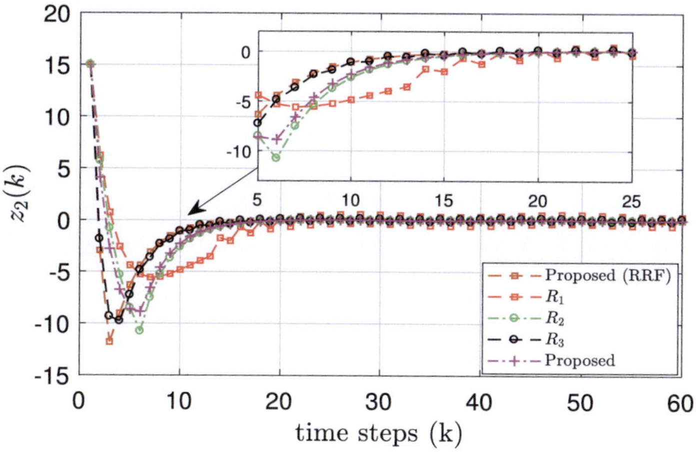

Fig. 5.11 State $z_2(k)$ of the perturbed pendulum system

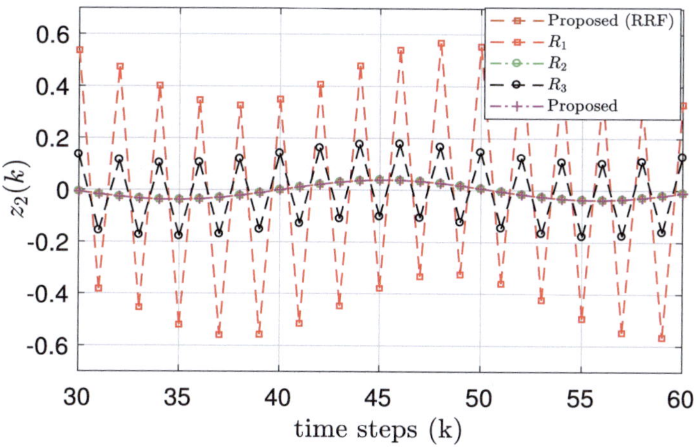

Fig. 5.12 State $z_2(k)$ during the sliding phase

reference trajectory following method to mitigate the disturbance effects, leading to improved transient and steady-state properties compared to [4], the proposed non-switching methods avoid these oscillations altogether.

Additionally, [5], the proposed (without RRF), and the proposed (RRF) methods, being based on non-switching control strategies, do not exhibit oscillatory behavior in the steady state, as indicated by their smoother trajectories. This trend is similarly observed in the angular velocity of the pendulum, depicted in Figs. 5.11 and 5.12.

The evolution of the sliding variable is illustrated in Figs. 5.13 and 5.14. The sliding variable with the proposed law converges to the ultimate band more quickly and without producing unnecessary oscillations compared to the other methods. Since $|c^{\top}b| = 0.027$, which is less than 1, the ultimate band width is bounded by $|s(k)| \leq$

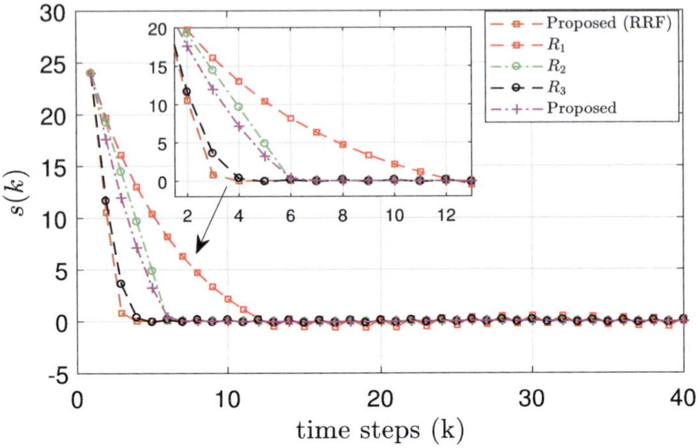

Fig. 5.13 Trajectory of the switching variable in the perturbed pendulum system

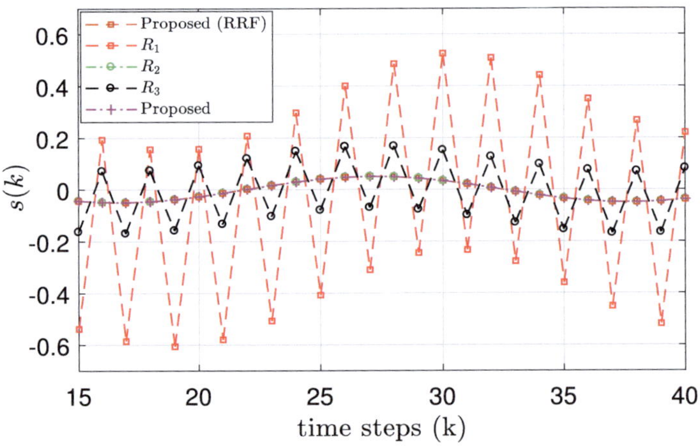

Fig. 5.14 Steady-state bands of the switching variable for various reaching laws

0.027. Furthermore, despite keeping γ and ϖ identical for both the proposed (without RRF) and proposed (RRF) methods, the latter demonstrates better performance due to the inclusion of the ϱ parameter, as illustrated graphically. The ultimate band width is notably larger in the cases of [4] and [6], which exhibit more significant oscillations.

For the methods [5], proposed (without RRF), and proposed (RRF), we found no oscillatory behaviour, and the size of the ultimate band is smaller. Finally, the control input can be observed in Figs. 5.15 and 5.16, where similar observations are made regarding the control signals. Nevertheless, it may be difficult to clearly distinguish the control signals from different reaching laws due to the congestion of data points, we can observe the chattering behavior, which is significant in [4] and [6], as seen

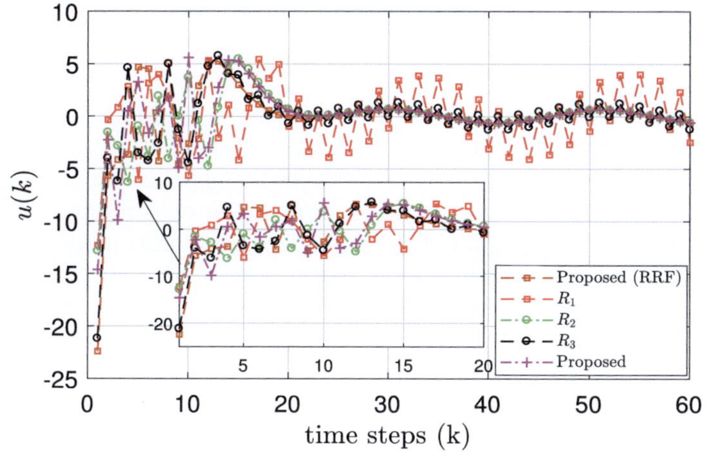

Fig. 5.15 Control signal dynamics for the perturbed pendulum system

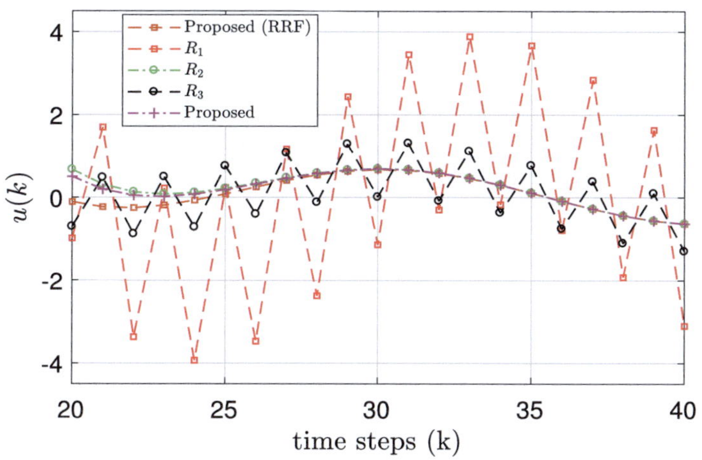

Fig. 5.16 Steady-state control signal

in Fig. 5.16. In contrast, the rest of the three approaches, which are non-switching based, do not exhibit chattering.

5.5 Conclusion

In this chapter, a novel reaching law utilizing difference equation with a minimum function, enhanced by a rate-regulatory function, was discussed for discrete sliding mode control. This modified approach solves the drawbacks of both Gao's reaching law and Utkin's method by effectively steering the switching variable's evolution. The discussed reaching law successfully mitigates chattering and reduces the need for excessive control efforts by introducing a design parameter, denoted as ϱ. Importantly, when the plant trajectories are farther from the equilibrium, the reaching law directs the switching variable towards the manifold while controlling its rate of change to stay within a manageable limit. This introduces a balance between control effort and system response time, allowing for tuning this balance by appropriately selecting the parameter ϱ. Simulation results depicts the effectiveness of the discussed method, and a comparative study using a pendulum system shows that the method enables fast convergence without excessive rate of change in the switching variable. Furthermore, it leads to a smaller steady-state bandwidth without generating unwanted oscillations.

In the next chapter, we will focus on the Lyapunov characterization of the discrete sliding mode control method introduced in this study.

References

1. Prasun, P., Kamal, S., Bartoszewicz, A., Ghosh, S.: Difference equation with minima based discrete-time sliding mode control. IEEE Trans. Circ. Syst. II: Expr. Briefs **70**(12), 4404–4408 (2023)
2. Prasun, P., Kamal, S., Ghosh, S., Dinh, T.N.: A minimum operator based discrete variable structure control. European J. Control **76**, 100953 (2024)
3. Bartoszewicz, A., Lesniewski, P.: New switching and nonswitching type reaching laws for SMC of discrete time systems. IEEE Trans. Control Syst. Tech. **24**(2), 670–677 (2016)
4. Gao, W., Wang, Y., Homaifa, A.: Discrete-time variable structure control systems. IEEE Trans. Ind. Electron. **42**(2), 117–122 (1995)
5. Bartoszewicz, A.: Discrete-time quasi-sliding-mode control strategies. IEEE Trans. Ind. Electron. **45**(4), 633–637 (1998)
6. Bartoszewicz, A., Adamiak, K.: Discrete-time sliding-mode control with a desired switching variable generator. IEEE Trans. Autom. Control **65**(4), 1807–1814 (2020)

Chapter 6
A Lyapunov Characterization of Difference Equation with Minima Based Discrete Sliding Mode Control

Abstract This chapter presents a Lyapunov-based analysis of a difference equation utilizing a minima-based law for discrete-time sliding mode control. We derive Lyapunov stability conditions and demonstrate the finite-time input-to-state stability of perturbed discrete-time systems. The analysis shows that the proposed method strikes a balance between Gao's and Utkin's reaching laws, effectively mitigating the key limitations of each–namely, chattering in Gao's approach and overly large control actions in Utkin's. The minima-based difference equation is applied to the design of discrete-time sliding mode control for both first- and second-order perturbed systems. Simulations are provided to compare the proposed method with existing discrete sliding mode control (DSMC) approaches, highlighting its superior performance.

6.1 Introduction

In this dedicated chapter, a comprehensive exploration of the Lyapunov characterization for the discrete-time sliding mode control is undertaken [1]. This analysis focuses on the difference equation with a minima-based law (DEM), with the goal of establishing Lyapunov stability conditions and demonstrating finite-time input-to-state stability for perturbed discrete-time systems. Through these derived conditions, we unveil the distinctive features of the proposed method, strategically positioned between Gao's reaching law and Utkin's reaching law, effectively mitigating the inherent limitations of both, namely chattering in the former and excessive control action in the latter. The application of the difference equation with minima-based discrete-time sliding mode control is extended to handle both first-order and second-order perturbed systems. To substantiate the effectiveness of this approach, a series of simulations are conducted, presenting a comparative study against existing DSMC design methods and showcasing the superior performance of the proposed methodology.

S. Kamal and P. Prasun, *A Minimum Function-Based Discrete Sliding Mode Control: Theory and Applications*, Studies in Systems, Decision and Control 621,
https://doi.org/10.1007/978-981-95-1161-7_6

The structure of this chapter is outlined as follows. Section 6.2 presents the preliminaries that lay the foundation for obtaining the main results. In Sect. 6.3, the main results of the study are presented. Section 6.4 focuses on validating the effectiveness of the proposed scheme through simulations and experiments. Finally, in Sect. 5.5, the chapter concludes the work.

6.2 Preliminaries

In this section, we revisit some the background results which will be helpful for comprehensive understanding of this chapter.

The following inequality will be useful while deriving the key results.

If $|a| \geq |b|$, with either $a, b \in \mathbb{R}_+$, or $a, b \in \mathbb{R}_-$, then

$$|a - b| \leq |a| - |b| \tag{6.1}$$

6.2.1 *Discrete-Time First-Order Unperturbed System*

Here, the finite-time stabilization of first-order unperturbed discrete-time systems [2] is revisited.

Consider the discrete-time system given by:

$$z(k+1) = u(k), \;\; z(0) = z_0 \tag{6.2}$$

where $z \in D \subset \mathbb{R}$ represents the state vector of the system, $k \in \mathbb{Z}_{\geq 0}$ is the time index, and $u \in \mathbb{R}$ denotes the control signal. The control law is designed using the difference equation with a minima-based approach as:

$$u(k) = z(k) - \text{sign}(z(k)) \min(|z(k)|, l) \tag{6.3}$$

where $\text{sign}(z) = \frac{z}{|z|}$ for $z \neq 0$ and $\text{sign}(0) = 0$ with $l > 0$. Substituting (6.3) into (6.2), the closed-loop system is given by:

$$z(k+1) = z(k) - \text{sign}(z(k)) \min(|z(k)|, l).$$

This system exhibits finite-time stability as discussed in [2], with the settling time function expressed as $\mathcal{K}(z_0) = \left\lceil \frac{|z_0|}{l} \right\rceil$, where $\mathcal{K}(z_0)$ demonstrates lower-semicontinuity at each point $z \in D$. The inclusion of $\min(|z(k)|, l)$ ensures that the system has a reaching phase for $|z(k)| > l$ and a dead-beat response for $|z(k)| \leq l$.

In the next section, we extend these results to the perturbed case within the framework of DSMC.

6.3 Lyapunov Characterization of DEM Based DSMC

Here, a new reaching law, which is based on difference equation with minima is presented. The stability conditions for such a law are developed by using Lyapunov methods. Let us go through the development process in the upcoming subsections.

6.3.1 *Discrete-Time First-Order Perturbed System*

Let us consider the following discrete-time system with perturbation as

$$z(k+1) = u(k) + d(k),\ z(0) = z_0 \tag{6.4}$$

where $z \in D \subset \mathbb{R}, k \in \mathbb{Z}_{\geq 0}, u \in \mathbb{R}$ and d is the matched type of bounded perturbation satisfying $|d| \leq d_0$. We design the control law as in (6.3), which yields the closed-loop system as $z(k+1) = z(k) - \text{sign}(z(k)) \min(|z(k)|, l) + d(k)$, where $l > d_0$. The closed-loop dynamics are FTISS in presence of the disturbance $d(k)$ and the solution $z(k)$ is ultimately bounded by d_0. We state the following theorem to formalize these claims.

Theorem 6.1 *We consider the dynamical system* (6.4) *with the control law* (6.3). *Assume there exists a continuous function* $V : D \to \mathbb{R}_+$ *that satisfies the following:*

$$\alpha_1(z(k)) \leq V(z) \leq \alpha_2(z(k)), \forall z \in D \backslash \{0\}, \tag{6.5}$$

$$V(z = 0) = 0, \tag{6.6}$$

$$\Delta V(z) \leq -\min(V(z), l) + \alpha_3(|d(k)|) \tag{6.7}$$

where $\alpha_1, \alpha_2, \alpha_3 \in$ *class* $\mathcal{K}$ *function. Then the system* (6.4) *is FTISS and* $z(k)$ *is ultimately bounded by* d_0*. Furthermore, the expression for the settling time is given by*

$$\mathcal{K}(d_0, z_0) \leq \left\lceil \frac{V(z_0) - d_0}{l - d_0} \right\rceil, \quad z_0 \in D \setminus \mathcal{B}_{d_0}(0),\ l > d_0,$$

where $\mathcal{K}(d_0, z_0)$ *exhibits the lower-semicontinuity property on D.*

Proof We consider the case where $|z_0| > l$, since for $|z_0| < l$, the solution converges to $z(k) = d(k)$ in finite time, with $\mathcal{K}(d_0, z_0) = 1$. To proceed with the analysis, we choose a Lyapunov candidate function $V(z(k)) = |z(k)|$. Then,

$$\begin{aligned}\Delta V(z) =& V(z(k+1)) - V(z(k)) \\ =& |z(k) - \text{sign}(z(k)) \min(|z(k)|, l) + d(k)| - |z(k)|.\end{aligned}$$

Since, $|z(k)| \geq \min(|z(k)|, l)$, then using (6.1),

$$
\begin{aligned}
\Delta V(z) \leq& |z(k)| - |\min(|z(k)|, l)| + d_0 - |z(k)| \\
=& - |\min(|z(k)|, l)| + d_0 \\
=& - \min(V(z), l) + d_0.
\end{aligned} \tag{6.8}
$$

This implies that the system (6.4) with the reaching law (6.3) exhibits finite-time input-to-state stability (FTISS) with $|z(k)| \leq d_0$, for all $k \geq \left\lceil \frac{V(z_0)-d_0}{l-d_0} \right\rceil$. The settling time expression is given by

$$
\mathcal{K}(d_0, z_0) \leq \left\lceil \frac{V(z_0) - d_0}{l - d_0} \right\rceil, \quad z_0 \in D \setminus \mathcal{B}_{d_0}(0),\ l > d_0.
$$

This settling time expression can be derived from (6.8), as shown below. Rewriting (6.8) for $V(z(k)) > l$, we get:

$$
V(k+1) \leq V(k) - l + d_0.
$$

Furthermore, we can express

$$
\begin{aligned}
V(k) \leq& V(k-1) - (l - d_0) \\
\leq& V(k-2) - 2(l - d_0) \\
&\vdots \\
\leq& V(0) - k(l - d_0).
\end{aligned} \tag{6.9}
$$

If at the kth time step, $V(0) - k(l - d_0) \leq l$, then from (6.9), $V(z(k)) \leq l$, which implies $V(k+1) \leq d_0$ for all $k \geq \left\lceil \frac{V(z_0)-d_0}{l-d_0} \right\rceil$. This completes the proof.

Remark 6.1 Note that in the system (6.4), if there is no perturbation, i.e., $d(k) = 0$, the closed-loop system exhibits finite-time stability with $\Delta V(z) \leq -\min(V(z), l)$. Furthermore, the expression for the settling time reduces to

$$
\mathcal{K}(z_0) = \left\lceil \frac{V(z_0)}{l} \right\rceil.
$$

6.3.2 Discrete-Time Second-Order Perturbed System

Let us consider the discrete-time dynamical system given by:

$$
\begin{aligned}
z_1(k+1) =& z_2(k) \\
z_2(k+1) =& u(k) + d(k)
\end{aligned} \tag{6.10}
$$

where $z = [z_1, z_2]^\top \in D \subset \mathbb{R}^2$ represents the state vector of the system, $k \in \mathbb{Z}_{\geq 0}$, $u \in \mathbb{R}$ denotes the control signal, and d is the disturbance satisfying $|d| \leq d_0$.

We define a switching function as $s(k) = c\, z_1(k) + z_2(k)$, where $-1 < c < 1$ is a positive scalar. The dynamics of the switching variable are given by $s(k+1) = c\, z_1(k+1) + z_2(k+1)$. Using (6.10), we can express this as $s(k+1) = c\, z_2(k) + u(k) + d(k)$.

The difference equation with the minima-based control input $u(k)$ is designed as

$$u(k) = -c\, z_2(k) + \mu_1 s(k) - \text{sign}(s(k)) \min(\mu_1 |s(k)|, l_1) \tag{6.11}$$

where $l_1 > d_0$, and $\mu_1 \in (0, 1)$. Substituting (6.11) into the sliding variable dynamics, we obtain the closed-loop system dynamics as

$$z_1(k+1) = s(k) - c\, z_1(k), \tag{6.12}$$

$$s(k+1) = \mu_1 s(k) - \text{sign}(s(k)) \min(\mu_1 |s(k)|, l_1) + d(k). \tag{6.13}$$

The dynamics of the switching variable exhibit finite-time input-to-state stability (FTISS), ensuring that $s(k)$ is ultimately bounded by d_0. Furthermore, the system states converge asymptotically to an invariant set $\omega = \{z \in D \mid |s(k)| \leq d_0\}$, and once within this set, they remain there for all future time. To formalize these claims, we propose the following theorem.

Theorem 6.2 *We consider the dynamical system* (6.12)–(6.13). *Let there exist continuous functions $V : D \to \mathbb{R}_+$ and $V_1 : D \to \mathbb{R}_+$, which satisfy the following:*

$$\alpha_1(s(k)) \leq V_1(s) \leq \alpha_2(s(k)), \forall s \in D \backslash \{0\}, \tag{6.14}$$

$$V_1(s = 0) = 0, \tag{6.15}$$

$$\Delta V_1(s) \leq -\min(V_1(s), l_1) + \alpha_3(|d(k)|) \tag{6.16}$$

where $\alpha_1, \alpha_2, \alpha_3 \in$ class $\mathcal{K}$ functions. Then, the sliding variable dynamics (6.13) *are FTISS, and $s(k)$ is ultimately bounded by d_0. Furthermore, the expression for the settling time is given by*

$$\mathcal{K}(d_0, s_0) \leq \left\lceil \frac{V(s_0) - d_0}{l_1 - d_0} \right\rceil, \quad s_0 \in D \setminus \mathcal{B}_{d_0}(0), \quad l_1 > d_0,$$

where $\mathcal{K}(d_0, s_0)$ is lower-semicontinuous on D. Moreover, if $V(z_1, s) > 0$ for all $(z_1, s) \in D \setminus \{0\}$, $V(0, 0) = 0$, and $\Delta V(z_1, s) \leq 0$ for all $|z_1| \geq \frac{\mu_1 \gamma - l_1 + d_0}{1 - |c|}$ and $|s| \leq \gamma$, then the closed-loop system (6.12)–(6.13) *is Lyapunov stable.*

Proof By performing a similar analysis as done for the first-order systems, we can show that the dynamics in (6.13) are FTISS, and that $s(k)$ is ultimately bounded by d_0. For this, by considering the Lyapunov function $V_1(s) = |s(k)|$, similar results can be obtained (analogous to the proof of Theorem 1), with the settling time function given by

$$\mathcal{K}(d_0, s_0) \leq \left\lceil \frac{V(s_0) - d_0}{l_1 - d_0} \right\rceil, \quad s_0 \in D \setminus \mathcal{B}_{d_0}(0), \quad l_1 > d_0.$$

Moreover, the condition $\Delta V_1(s) \leq 0$ within the set $|s| \leq \gamma$ ensures its positive invariance. Now, we will show that the closed-loop system in (6.12)–(6.13) is Lyapunov stable. Consider the Lyapunov function $V(k) = |z_1(k)| + |s(k)|$. Then,

$$\begin{aligned}\Delta V(k) =& V(k+1) - V(k)\\ =& |z_1(k+1)| + |s(k+1)| - |z_1(k)| - |s(k)|\\ =& |s(k) - cz_1(k)| + |\mu_1 s(k) - \operatorname{sign}(s(k)) \min(\mu_1|s(k)|, l_1)\\ &+ d(k)| - |z_1(k)| - |s(k)|.\end{aligned}$$

Since, $\mu_1|s(k)| \geq \min(\mu_1|s(k)|, l_1)$, then using inequality (6.1),

$$\begin{aligned}\Delta V(k) \leq& |s(k)| + |c||z_1(k)| + \mu_1|s(k)| - |\min(\mu_1|s(k)|, l_1)| + d_0\\ &- |z_1(k)| - |s(k)|.\end{aligned}$$

Since, $l_1 < |s(k)| \leq \gamma$, it further follows

$$\begin{aligned}\Delta V(k) \leq& \mu_1\gamma + |c||z_1(k)| - l_1 + d_0 - |z_1(k)|\\ \leq& 0, \ \forall\, |z_1(k)| \geq \frac{\mu_1\gamma - l_1 + d_0}{1 - |c|}.\end{aligned} \tag{6.17}$$

Thus, if $|z_1(0)| \leq \frac{\mu_1\gamma - l_1 + d_0}{1-|c|}$, then $|z_1(k)| \leq \frac{\mu_1\gamma - l_1 + d_0}{1-|c|}$ for all $k \geq 0$, and the set

$$\Omega = \left\{ |z_1| \leq \frac{\mu_1\gamma - l_1 + d_0}{1 - |c|}, \ |s| \leq \gamma \right\}$$

is positively invariant if $l_1 \geq d_0$. Moreover, any solution that starts in Ω will reach the set

$$\omega = \{z \in \Omega \subset D \mid |s(k)| \leq d_0\}$$

within the settling time $\mathcal{K}(d_0, s_0)$. Hence, the closed-loop system in (6.12)–(6.13) is Lyapunov stable. This completes the proof. □

6.3.3 *Backstepping-Based Sliding Mode Control Design for Discrete-Time Second-Order Systems*

In this subsection, we present the backstepping approach for designing a difference equation with a minima-based control law, ensuring that all states converge to an invariant set in finite time. Consider the following dynamical system:

$$\begin{aligned} z_1(k+1) =& z_2(k) \\ z_2(k+1) =& u(k) + d(k) \end{aligned} \tag{6.18}$$

where $z = [z_1, z_2]^\top \in D \subset \mathbb{R}^2$ is the system state vector, $k \in \mathbb{Z}_{\geq 0}$, $u \in \mathbb{R}$ is the control input, and d is the disturbance satisfying $|d| \leq d_0$.

We define a nonlinear sliding surface as

$$s(k) = z_2(k) - \phi(z_1(k)),$$

where $\phi(z_1(k)) := \mu_1 z_1(k) - \text{sign}(z_1(k)) \min(\mu_1 |z_1(k)|, l_1)$, with $\mu_1 \in (0, 1)$ and $l_1 \in \mathbb{R}_+$.

Next, the control input is designed as

$$u(k) = \phi(z_1(k+1)) + \mu_2 s(k) - \text{sign}(s(k)) \min(\mu_2 |s(k)|, l_2),$$

where $\mu_2 \in (0, 1)$ and $l_2 \in \mathbb{R}_+$. With this control law, the closed-loop system becomes

$$z_1(k+1) = s(k) + \phi(z_1(k)), \tag{6.19}$$

$$s(k+1) = \mu_2 s(k) - \text{sign}(s(k)) \min(\mu_2 |s(k)|, l_2) + d(k). \tag{6.20}$$

The closed-loop system (6.19)–(6.20) is FTISS, with the system states converging to an invariant set within a finite time. To justify these claims, we propose the following theorem.

Theorem 6.3 *Let us consider the dynamical system* (6.19)–(6.20). *We assume that there exist a continuous function $V : D \to \mathbb{R}_+$ that satisfies the following:*

$$\alpha_1(z(k)) \leq V(z) \leq \alpha_2(z(k)), \forall z \in D \backslash \{0\}, \tag{6.21}$$

$$V(z = 0) = 0, \tag{6.22}$$

$$\Delta V(z) \leq -\min(V(z), l) + \alpha_3(|d(k)|) \tag{6.23}$$

where $\alpha_1, \alpha_2, \alpha_3 \in$ class $\mathcal{K}$ functions. Then, the system (6.19)–(6.20) *is FTISS, and the sliding variable is ultimately bounded by d_0. Furthermore, the settling time function is given by*

$$\mathcal{K}(d_0, z_0) \leq \left\lceil \frac{V(z_0) - 2d_0}{l - 2d_0} \right\rceil, \quad z_0 \in D \setminus \mathcal{B}_{2d_0}(0),$$

where $l = \min(l_1, l_2) > 2d_0$, and $\mathcal{K}(d_0, z_0)$ is lower semicontinuous on D.

Proof For the analysis, we select a Lyapunov candidate function as $V(z(k)) = |z_1(k)| + |s(k)|$. Then,

$$
\begin{aligned}
\Delta V(z) &= V(z(k+1)) - V(z(k)) \\
&= |s(k) + \phi(z_1(k))| + |\mu_2 s(k) - \text{sign}(s(k)) \min(\mu_2|s(k)|, l_2) \\
&\quad + d(k)| - |z_1(k)| - |s(k)|.
\end{aligned}
$$

Since $\mu_1|z_1(k)| \geq \min(\mu_1|z_1(k)|, l_1)$ and $\mu_2|s(k)| \geq \min(\mu_2|s(k)|, l_2)$, by using inequality (6.1),

$$
\begin{aligned}
\Delta V(z) &\leq |s(k)| + \mu_1|z_1(k)| - \min(\mu_1|z_1(k)|, l_1) + \mu_2|s(k)| \\
&\quad - \min(\mu_2|s(k)|, l_2) + d_0 - |z_1(k)| - |s(k)| \\
&= (\mu_1 - 1)|z_1(k)| + \mu_2|s(k)| - \min(\mu_1|z_1(k)|, l_1) \\
&\quad - \min(\mu_2|s(k)|, l_2) + d_0.
\end{aligned} \tag{6.24}
$$

Based on the result of $\min(\cdot, \cdot)$, inequality (6.24) is further derived. This leads to four scenarios that cover the entire admissible state space.

Scenario I: $\mu_1|z_1(k)| \leq l_1$ and $\mu_2|s(k)| > l_2$
Under these conditions, inequality (6.24) can be written as

$$
\Delta V(z) \leq (\mu_1 - 1)|z_1(k)| + \mu_2|s(k)| - \mu_1|z_1(k)| - l_2 + d_0.
$$

Since $\mu_1|z_1(k)| \leq l_1$ implies $\phi(z_1(k)) = 0$, it follows immediately that $z_1(k) = s(k)$ (from (6.19)). Thus, the inequality can be further rewritten as

$$
\begin{aligned}
\Delta V(z) &\leq -(1 - \mu_2)|s(k)| - l_2 + d_0 \\
&\leq -l_2 + d_0.
\end{aligned} \tag{6.25}
$$

Scenario II: $\mu_1|z_1(k)| > l_1$ and $\mu_2|s(k)| \leq l_2$
Inequality (6.24) simplifies to

$$
\Delta V(z) \leq (\mu_1 - 1)|z_1(k)| + \mu_2|s(k)| - l_1 - \mu_2|s(k)| + d_0.
$$

Further, it simplifies to

$$
\Delta V(z) \leq -l_1 + d_0. \tag{6.26}
$$

Scenario III: $\mu_1|z_1(k)| > l_1$ and $\mu_2|s(k)| > l_2$
In this scenario, inequality (6.24) simplifies to

$$
\Delta V(z) \leq (\mu_1 - 1)|z_1(k)| + \mu_2|s(k)| - l_1 - l_2 + d_0.
$$

Furthermore, since at steady state $|s(k)| \leq d_0$, we can rewrite the inequality as

$$
\Delta V(z) \leq -(l_1 + l_2) + 2d_0. \tag{6.27}
$$

Scenario IV: $\mu_1|z_1(k)| \leq l_1$ and $\mu_2|s(k)| \leq l_2$
In this case, inequality (6.24) simplifies to

$$\Delta V(z) \leq (\mu_1 - 1)|z_1(k)| + \mu_2|s(k)| - \mu_1|z_1(k)| - \mu_2|s(k)| + d_0,$$

which reduces to

$$\Delta V(z) \leq -(V(z(k)) - |s(k)|) + d_0.$$

Furthermore, under this scenario, the governing dynamics in (6.20) imply that $|s(k)| \leq d_0$, and thus, we can write

$$\Delta V(z) \leq -V(z(k)) + 2d_0. \tag{6.28}$$

Moreover, if $s(0) \neq 0$, the system dynamics in (6.19)–(6.20) will continue to evolve, and *Scenario IV* will transition into one of the other three scenarios. Therefore, combining (6.25), (6.26), (6.27), and (6.28), we can express the following:

$$\Delta V(z) \leq -\min(V(z(k)), l) + 2d_0. \tag{6.29}$$

In this case, the system states are designed to converge to an invariant set within a finite time. Furthermore, the computation of the settling time follows a similar process to that in Theorem 1. Thus, the settling time function is given by

$$\mathcal{K}(d_0, z_0) \leq \left\lceil \frac{V(z_0) - 2d_0}{l - 2d_0} \right\rceil, \quad z_0 \in D \setminus \mathcal{B}_{2d_0}(0), \quad l > 2d_0.$$

This completes the proof. □

6.4 Simulation Examples

In this section, we analyze perturbed discrete-time systems of both first-order and second-order. We design a difference equation with a minima-based control law for these systems and verify the effectiveness of the proposed control law.

6.4.1 First-Order System

Consider the perturbed system as given in (6.4). For this system, we introduce a sinusoidal disturbance defined by $d(k) = \sin(0.6k)$. To control the system, we design a difference equation with a minima-based reaching control law, as described in (6.3).

The system starts with the initial condition $z(0) = 100$, and the control gain is selected as $l = 6$. The evolution of the system state under the influence of the bounded

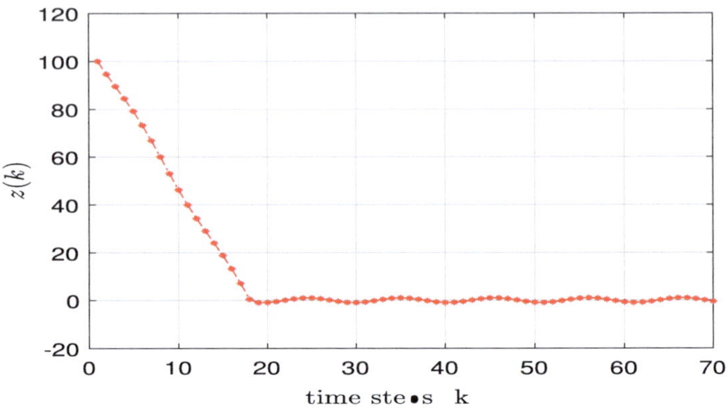

Fig. 6.1 The state $z(k)$ of the perturbed system

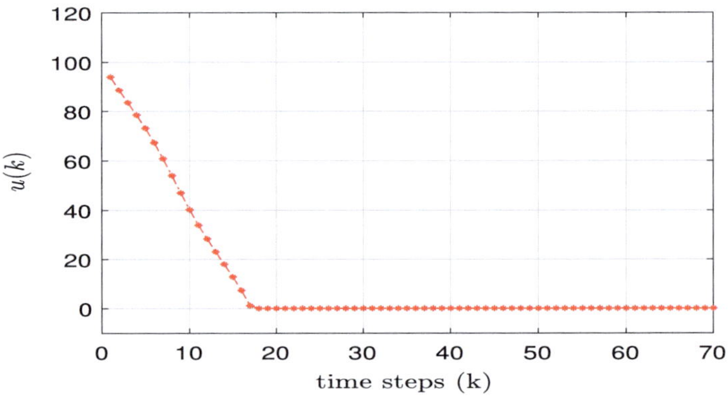

Fig. 6.2 The control input $u(k)$ for the perturbed system

disturbance is shown in Fig. 6.1. As seen in the plot, after a finite number of time steps, the system state converges to an invariant set and remains within this set for all future time steps.

Additionally, the evolution of the control input is shown in Fig. 6.2. This plot demonstrates that the control input does not exhibit any chattering, confirming the effectiveness and smoothness of the proposed control law.

6.4.2 Second-Order System: A Comparative Analysis

In this section, the proposed method is compared with the reaching laws presented in Chakrabarty et al. [3] and Bartoszewicz et al. [4]. We now consider the following unstable perturbed dynamical system:

$$\begin{aligned} z_1(k+1) &= z_2(k), \\ z_2(k+1) &= 2z_1(k) - 0.3z_2(k) + u(k) + d(k). \end{aligned} \tag{6.30}$$

We consider a sinusoidal disturbance given by $d(k) = \sin(0.6k)$, with the disturbance magnitude $d_m = 1$. The initial conditions for the system are chosen as $z_1(0) = 100$ and $z_2(0) = 50$. For all the methods, the sliding surface parameter is selected as $c = 0.5$. The notations for different parameters are taken from the corresponding references.

As per the band approach analysis in Chakrabarty et al. [3], where $d_m \leq \delta \leq 2d_m$, the control law is designed by choosing the minimum possible band, which results in $\delta = d_m$. For this choice, the design parameters are computed as $\mu = 1$ and $\epsilon = -d_m$, which leads to Utkin's reaching law.

In the reference trajectory-based approach proposed in Bartoszewicz et al. [4], the design parameters are selected as $\sigma_0 = 30$ and $\epsilon = 1.1$. For the proposed method, the design parameters are chosen as $l_1 = 8$ and $\mu_1 = 0.8$.

The evolution of the system states is shown in Fig. 6.3. The system states driven by the method in [3] exhibit oscillatory behavior and a high rate of change in the initial phase. In contrast, the rate of change of the states for the methods in [4] and the proposed method is lower. However, oscillatory behavior persists during the sliding phase in the approach of [4], unlike in the approaches of [3] and the proposed method, where this is not observed.

Additionally, the evolution of the switching function and control signal is depicted in Fig. 6.4. Since the approach in [3] follows Utkin's law from the start, it reaches the manifold in just one step, resulting in a very high rate of change of the sliding variable and the highest control signal magnitude during the initial phase. In comparison, both [4] and the proposed method show a lower rate of change in the sliding variable and comparatively smaller control signal magnitudes.

Furthermore, chattering is observed in the approach of [4], which is absent in the methods of [3] and the proposed method. It is also noteworthy that the ultimate band size in [3] and the proposed method is d_m, whereas in [4], it is $d_m + \epsilon$.

Remark 6.2 The minimum ultimate band size guaranteed by [3] is d_m, but this is achieved at the expense of a large control action. In contrast, [4] ensures lower control effort in the initial phase, but results in chattering and a larger bandwidth of $d_m + \epsilon$. However, the proposed method guarantees a smaller bandwidth (d_m) without requiring a large control effort in the initial phase and effectively eliminates chattering.

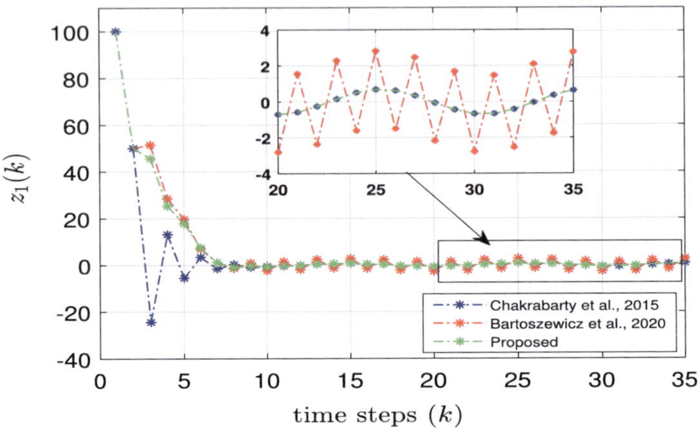

(a)

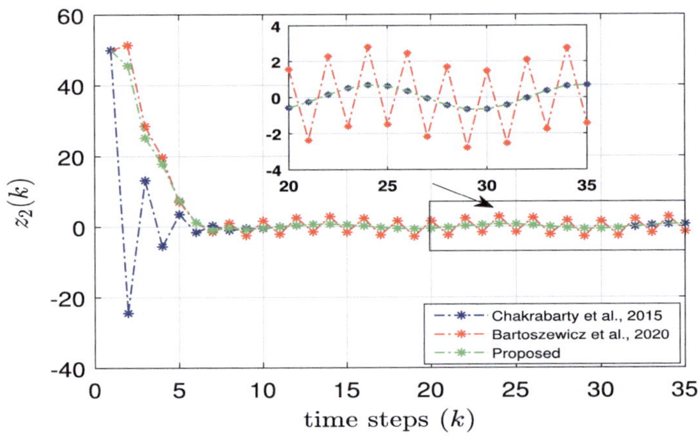

(b)

Fig. 6.3 The time evolution of the system states for the perturbed discrete-time system

6.5 Conclusion

In conclusion, this chapter presented a systematic Lyapunov characterization for the difference equation with minima-based discrete-time sliding mode control. The derived Lyapunov stability conditions not only validate the stability of the systems but also demonstrate the successful mitigation of the limitations present in both Gao's reaching law approach and Utkin's equivalent control-based method. The effectiveness of the proposed methodology was examined in the context of first-order and second-order perturbed discrete-time systems. To substantiate its performance, a comparative study, supported by simulations, was conducted, highlighting the robust performance of the DEM-based discrete-time sliding mode control.

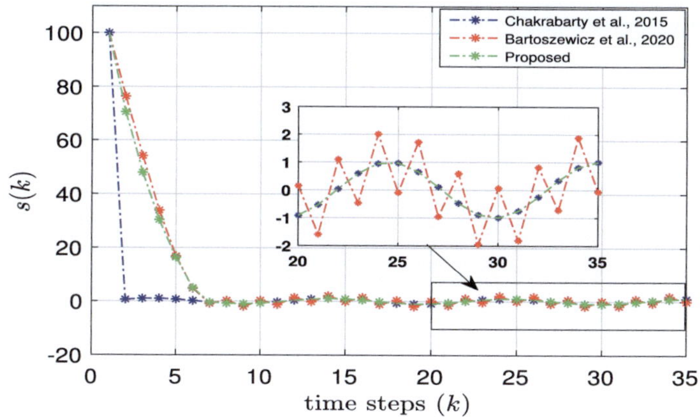

(a)

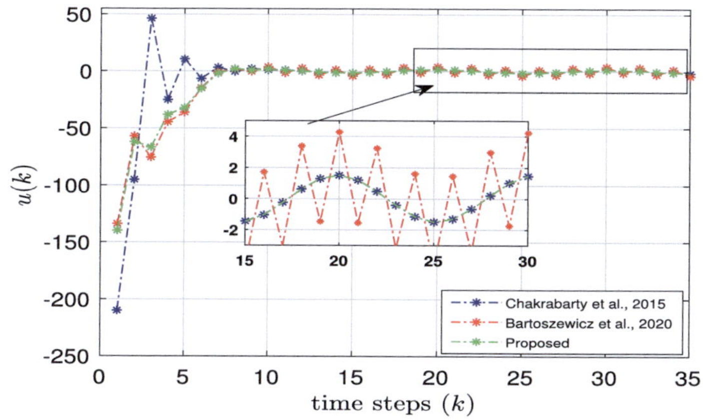

(b)

Fig. 6.4 The time evolution of the switching function and control signal for the perturbed discrete-time system

In the next chapter, we will introduce discrete-time twisting-like and super-twisting-like algorithms. We will provide an in-depth analysis of their stability using Lyapunov methods, extending the analytical framework established in this chapter.

References

1. Prasun, P., Kamal, S., Pandey, S., Bartoszewicz, A., Ghosh, S.: A minimum operator based discrete-time sliding mode control. IEEE Trans. Autom. Control **69**(11), 7871–7876 (2024)
2. Haddad, W.M., Lee, J.: Finite-time stability of discrete autonomous systems. Automatica **122** (2020)

3. Chakrabarty, S., Bandyopadhyay, B.: Minimum ultimate band design of discrete sliding mode control. Asian J. Control **17**(5), 1889–1897 (2015)
4. Bartoszewicz, A., Adamiak, K.: Discrete-time sliding-mode control with a desired switching variable generator. IEEE Trans. Autom. Control **65**(4), 1807–1814 (2020)

Chapter 7
DEM Based Discrete-Time Twisting-like Algorithm and Discrete-Time Super-Twisting-like Algorithm

Abstract In this chapter, we propose a difference equation with a minima-based approach for developing discrete-time twisting-like and super-twisting-like algorithms. In contrast to most existing methods that rely on Euler discretization, the proposed algorithms are characterized by distinct roots. These algorithms are designed to handle both unperturbed and perturbed scenarios. Furthermore, we investigate the use of the discrete-time super-twisting-like algorithm in the design of a discrete-time super-twisting observer (STO), which facilitates finite-time state estimation. To validate the effectiveness of this approach, we present a case study involving a perturbed pendulum system with both known and unknown inputs, focusing on the state observation problem. Simulation results demonstrate the effectiveness of the proposed algorithm.

7.1 Introduction

In this chapter, a novel approach based on the difference equation with minima is introduced for the design of discrete-time twisting-like and super-twisting-like algorithms [1, 2]. These algorithms differ from traditional methods, which typically rely on Euler discretization, and are developed for both unperturbed and perturbed scenarios. Additionally, the discrete-time super-twisting-like algorithm is applied to the design of a discrete-time super-twisting observer (STO) for finite-time state estimation. To demonstrate the effectiveness of the proposed method, an illustrative case study involving a perturbed pendulum system with both known and unknown inputs is presented, accompanied by simulation results.

The structure of this chapter is as follows: Sect. 7.2 introduces the discrete-time twisting-like algorithm, along with simulation results. Section 7.3 presents the results for the discrete-time super-twisting-like algorithm, along with corresponding simulation outcomes. In Sect. 7.4, the design of the discrete-time super-twisting observer is discussed, with its application to the state estimation problem of the perturbed pendulum system. Finally, Sect. 7.5 concludes the chapter.

S. Kamal and P. Prasun, *A Minimum Function-Based Discrete Sliding Mode Control: Theory and Applications*, Studies in Systems, Decision and Control 621,
https://doi.org/10.1007/978-981-95-1161-7_7

7.2 DEM Based Twisting-like Algorithm for Discrete-Time System

This section introduces a novel discrete-time twisting-like algorithm designed specifically for discrete-time systems [1]. The proposed algorithm effectively drives the system states to zero within a finite number of time steps, ensuring fast convergence. Notably, it avoids chattering, thus eliminating the need for excessive control effort.

Let us consider the following dynamical system:

$$\begin{aligned} z_1(k+1) =& \mu_1 z_2(k), \\ z_2(k+1) =& \mu_2 z_1(k) - \operatorname{sign}(z_1(k)) \min(\mu_2|z_1(k)|, l_1) + \mu_3 z_2(k) \\ & - \operatorname{sign}(z_2(k)) \min(\mu_3|z_2(k)|, l_2) \end{aligned} \tag{7.1}$$

where $z = [z_1, z_2]^\top \in D \subset \mathbb{R}^2$ represents the system state vector, $k \in \mathbb{Z}_{\geq 0}$, with $l_1 > 0, l_2 > 0$, and $\mu_1, \mu_2, \mu_3 \in (0, 1)$.

To establish the stability of this algorithm, we propose the following theorem.

Theorem 7.1 *Consider the dynamical system* (7.1) *with the condition* $\mu_1 + \mu_3 \leq 1$. *Let* $z = 0$ *denote the equilibrium point of* (7.1), *which lies within the domain* D. *Assume the existence of a continuous function* $V : D \to \mathbb{R}_+$ *that satisfies the following conditions:*

$$V(z) > 0, \quad \textit{for all} \;\; z \in D \backslash \{0\}, \tag{7.2}$$

$$V(z = 0) = 0, \tag{7.3}$$

$$\Delta V(z) \leq -\min(V(z), l), \tag{7.4}$$

where $l = \min(l_1, l_2)$. *Then, the solution* $z(k) \equiv 0$ *is finite-time stable. Furthermore, the settling time function is given by* $\mathcal{K}(z_0) \leq \lceil \frac{V(z_0)}{l} \rceil$, *for* $z_0 \in D$, *where* $\mathcal{K}(z_0)$ *is lower-semicontinuous on* D.

Proof. Consider the Lyapunov candidate function as $V(z(k)) = |z_1(k)| + |z_2(k)|$. Then, the Lyapunov function at the next time step is given by $V(z(k+1)) = |z_1(k+1)| + |z_2(k+1)|$. The change in the Lyapunov energy is thus obtained as

$$\begin{aligned} \Delta V(z) =& V(z(k+1)) - V(z(k)) \\ =& |\mu_1 z_2(k)| + |\mu_2 z_1(k) - \operatorname{sign}(z_1(k)) \min(\mu_2|z_1(k)|, l_1) + \mu_3 z_2(k) \\ & - \operatorname{sign}(z_2(k)) \min(\mu_3|z_2(k)|, l_2)| - |z_1(k)| - |z_2(k)|. \end{aligned}$$

Since $\mu_2|z_1(k)| \geq \min(\mu_2|z_1(k)|, l_1)$ and $\mu_3|z_2(k)| \geq \min(\mu_3|z_2(k)|, l_2)$, it follows that

$$\begin{aligned} \Delta V(z) \leq& (\mu_2 - 1)|z_1(k)| + (\mu_1 + \mu_3 - 1)|z_2(k)| \\ & - |\min(\mu_2|z_1(k)|, l_1)| - |\min(\mu_3|z_2(k)|, l_2)|. \end{aligned} \tag{7.5}$$

Inequality (7.5) is further derived from the result of the function $\min(\cdot,\cdot)$. This leads to four distinct scenarios that cover the entire admissible state space.

Scenario I: $\mu_2|z_1(k)| \leq l_1$ and $\mu_3|z_2(k)| > l_2$

In this scenario, inequality (7.5) can be expressed as:

$$\Delta V(z) \leq (\mu_2 - 1)|z_1(k)| + (\mu_1 + \mu_3 - 1)|z_2(k)| - \mu_2|z_1(k)| - l_2$$

Since, $\mu_1 + \mu_3 \leq 1$, it further follows

$$\Delta V(z) \leq -|z_1(k)| - l_2 \leq -l_2. \tag{7.6}$$

Scenario II: $\mu_2|z_1(k)| > l_1$ and $\mu_3|z_2(k)| \leq l_2$

In this scenario, inequality (7.5) simplifies to:

$$\Delta V(z) \leq (\mu_2 - 1)|z_1(k)| + (\mu_1 + \mu_3 - 1)|z_2(k)| - l_1 - \mu_3|z_2(k)|$$

which further simplifies as

$$\begin{aligned}\Delta V(z) \leq& -(1 - \mu_2)|z_1(k)| - (1 - \mu_1)|z_2(k)| - l_1\\ \leq& - l_1.\end{aligned} \tag{7.7}$$

Scenario III: $\mu_2|z_1(k)| > l_1$ and $\mu_3|z_2(k)| > l_2$

In this case, inequality (7.5) simplifies to

$$\Delta V(z) \leq (\mu_2 - 1)|z_1(k)| + (\mu_1 + \mu_3 - 1)|z_2(k)| - l_1 - l_2,$$

which further reduces to:

$$\Delta V(z) \leq -l_1 - l_2 \leq -l. \tag{7.8}$$

Scenario IV: $\mu_2|z_1(k)| \leq l_1$ and $\mu_3|z_2(k)| \leq l_2$

In this case, inequality (7.5) simplifies to

$$\Delta V(z) \leq (\mu_2 - 1)|z_1(k)| + (\mu_1 + \mu_3 - 1)|z_2(k)| - \mu_2|z_1(k)| - \mu_3|z_2(k)|,$$

which further reduces to

$$\Delta V(z) \leq -(V(z(k)) - \mu_1|z_2(k)|).$$

Additionally, if $z_2(0) = 0$, then from (7.1), the inequality simplifies further to:

$$\Delta V(z) \leq -V(z(k)). \tag{7.9}$$

Note that the condition in (7.9) results in a dead-beat response, which would cause the entire system dynamics to collapse. Furthermore, if $z_2(0) \neq 0$, the governing dynamics in (7.1) will cause *Scenario IV* to evolve into one of the other scenarios. Finally, by combining the four conditions from (7.6), (7.7), (7.8), and (7.9), we obtain the following:

$$\Delta V(z) \leq -\min(V(z), l). \tag{7.10}$$

Furthermore, the calculation of the settling time function from (7.10) is straightforward and follows similar steps as demonstrated in the proof of Theorem 1. Therefore, the expression for the settling time function can be obtained as $\mathcal{K}(z_0) \leq \lceil \frac{V(z_0)}{l} \rceil$. This concludes the proof. □

Remark 7.1 The twisting-like algorithm performs effectively even in the presence of bounded perturbations. However, in this case, the states will remain within an invariant set rather than converging to zero. The Lyapunov-based condition for this scenario is given by $\Delta V(z) \leq -\min(V(z(k)), l) + 2d_0$, with the settling time function expressed as

$$\mathcal{K}(d_0, z_0) \leq \lceil \frac{V(z_0) - 2d_0}{l - 2d_0} \rceil, \quad z_0 \in D \backslash \mathcal{B}_{2d_0}(0),$$

where $l > 2d_0$.

7.2.1 *Illustrative Example*

Consider the dynamical system given by (7.1). The system is initialized with the state values $z_1(0) = 10$ and $z_2(0) = 15$. For the design of the control law, the following parameter values are selected: $l_1 = 2$, $l_2 = 2$, $\mu_1 = 0.5$, $\mu_2 = 0.5$, and $\mu_3 = 0.4$.

The trajectory of the system states, as shown in Fig. 7.1, illustrates how the states evolve over time. Since there is no perturbation acting on the system, both states converge to the origin in a finite number of time steps, which demonstrates the effectiveness of the algorithm in achieving finite-time stability.

To further visualize the system's behavior, Fig. 7.2 presents the phase portrait of the system. The phase portrait clearly shows the trajectories of the system states and their convergence towards the origin in finite time, confirming the theoretical analysis.

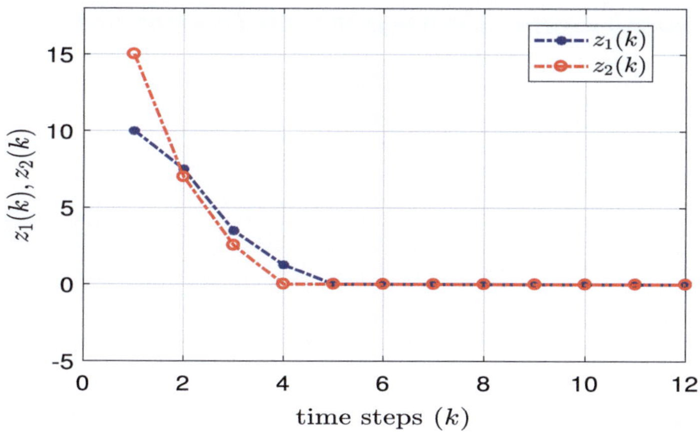

Fig. 7.1 Behavior of the system states over time

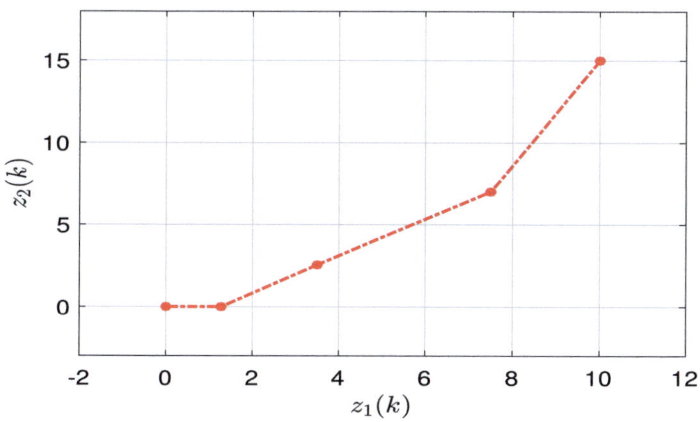

Fig. 7.2 Phase space evolution of the system

7.3 DEM Based Discrete-Time Super-Twisting-like Algorithm

In this section, we present a novel variant of the super-twisting-like algorithm, developed within the framework of difference equations with minima, specifically designed for discrete-time systems [2]. Unlike the Euler-discretized super-twisting algorithms commonly found in the literature, the proposed algorithm is distinguished by its remarkable ability to completely eliminate chattering.

7.3.1 Super-Twisting-like Algorithm for Unperturbed Discrete-Time Systems

Consider the nonlinear discrete-time dynamics given by

$$\begin{aligned} z_1(k+1) =& \wp_1 z_1(k) - l_1 \text{sign}(z_1(k)) \min\left(\frac{\wp_1 |z_1(k)|}{l_1}, |z_1(k)|^{\frac{1}{2}}\right) + z_2(k), \\ z_2(k+1) =& \wp_2 z_1(k) - \text{sign}(z_1(k)) \min\left(\wp_2 |z_1(k)|, l_2\right) \end{aligned} \tag{7.11}$$

where $z = [z_1, z_2]^\top \in D \subseteq \mathbb{R}^2$ represents the system states, $k \in \mathbb{Z}_{\geq 0}$, $\wp_1, \wp_2 \in (0, 1)$, and $l_1, l_2 \in \mathbb{R}_+$.

Remark 7.2 An important observation is that the proposed algorithm is not an Euler-discretization of the continuous-time Super-Twisting Algorithm (STA). Specifically, it is noteworthy that the second equation in (7.11) involves $z_1(k)$ rather than $z_2(k)$, which distinguishes it from the Euler-discretization-based STAs (see [3–6]).

The following theorem establishes the finite-time stability (FTS) of the zero solution of (7.11).

Theorem 7.2 *Consider the system* (7.11) *with the conditions* $\wp_1 + \wp_2 \leq 1$ *and* $l_1 > \left(1 + \left|\frac{z_2(k)}{z_1(k)}\right|\right)^{\frac{1}{2}}$, *where* $z_1(k) \neq 0$. *Let* $z = 0$ *represent the equilibrium point of* (7.11), *and let* $D \subset \mathbb{R}^2$ *be a domain that contains* $z = 0$. *Assume the existence of a continuous function* $V : D \to \mathbb{R}_{\geq 0}$ *that satisfies the following:*

$$V(z) > 0, \ \forall \, z \in D \backslash \{0\}, \tag{7.12}$$

$$V(z = 0) = 0 \text{ and} \tag{7.13}$$

$$\Delta V(z) \leq -\min(V(z), (V(z))^{\frac{1}{2}}, l_2), \ \forall \, z \in D \backslash \{0\}. \tag{7.14}$$

Then, the solution $z(k) \equiv 0$ *is finite-time stable. Furthermore, if* $D = \mathbb{R}^2$, $V(\cdot)$ *is radially unbounded, and the condition* $\Delta V(z) \leq -\min(V(z), (V(z))^{\frac{1}{2}}, l_2)$ *holds for all* $z \in \mathbb{R}^2 \backslash \{0\}$, *the solution* $z(k) \equiv 0$ *is globally finite-time stable.*

Proof: We examine the case when $z_1(k) \neq 0$, since if $z_1(k) = 0$, then from (7.11), we would have $z_1(k+1) = z_2(k)$ and $z_2(k+1) = 0$, which would lead to the system evolving further with $z_1(k) \neq 0$ until both $z_1(k)$ and $z_2(k)$ reach zero. To analyze this, we define the Lyapunov candidate function as $V(z(k)) = |z_1(k)| + |z_2(k)|$. Consequently, we have $V(z(k+1)) = |z_1(k+1)| + |z_2(k+1)|$. The change in Lyapunov energy is then given by

$$\begin{aligned} \Delta V(z) =& V(z(k+1)) - V(z(k)) \\ =& \left|\wp_1 z_1(k) - l_1 \text{sign}(z_1(k)) \min\left(\frac{\wp_1 |z_1(k)|}{l_1}, |z_1(k)|^{\frac{1}{2}}\right) + z_2(k)\right| \\ &+ \left|\wp_2 z_1(k) - \text{sign}(z_1(k)) \min\left(\wp_2 |z_1(k)|, l_2\right)\right| - |z_1(k)| - |z_2(k)| \end{aligned}$$

Since $\wp_1|z_1(k)| \geq l_1 \min\left(\frac{\wp_1|z_1(k)|}{l_1}, |z_1(k)|^{\frac{1}{2}}\right)$ and $\wp_2|z_1(k)| \geq \min(\wp_2|z_1(k)|, l_2)$, we can further express the inequality as

$$\Delta V(z) \leq -\left|l_1 \min\left(\frac{\wp_1|z_1(k)|}{l_1}, |z_1(k)|^{\frac{1}{2}}\right)\right| - \left|\min(\wp_2|z_1(k)|, l_2)\right| + (\wp_1 + \wp_2 - 1)|z_1(k)| \tag{7.15}$$

We analyze four possible cases that emerge from the system (7.11), covering the entire state space. These cases are determined based on the function obtained as the output of the minimum function. The four cases are as follows:

(i) If $\frac{\wp_1|z_1(k)|}{l_1} \leq |z_1(k)|^{\frac{1}{2}}$ and $\wp_2|z_1(k)| > l_2$,

(ii) If $\frac{\wp_1|z_1(k)|}{l_1} > |z_1(k)|^{\frac{1}{2}}$ and $\wp_2|z_1(k)| \leq l_2$,

(iii) If $\frac{\wp_1|z_1(k)|}{l_1} > |z_1(k)|^{\frac{1}{2}}$ and $\wp_2|z_1(k)| > l_2$,

(iv) If $\frac{\wp_1|z_1(k)|}{l_1} \leq |z_1(k)|^{\frac{1}{2}}$ and $\wp_2|z_1(k)| \leq l_2$.

These cases are analyzed, and the difference inequality (7.15) is subsequently derived.

Case I: If $\frac{\wp_1|z_1(k)|}{l_1} \leq |z_1(k)|^{\frac{1}{2}}$ and $\wp_2|z_1(k)| > l_2$.
Given that $\wp_1 + \wp_2 \leq 1$, we can further express the change in the Lyapunov function as:

$$\Delta V(z) \leq -l_1 \min\left(\frac{\wp_1|z_1(k)|}{l_1}, |z_1(k)|^{\frac{1}{2}}\right) - \min(\wp_2|z_1(k)|, l_2).$$

It follows that:

$$\Delta V(z) \leq -\wp_1|z_1(k)| - l_2 \leq -l_2 \tag{7.16}$$

Case II: If $\frac{\wp_1|z_1(k)|}{l_1} > |z_1(k)|^{\frac{1}{2}}$ and $\wp_2|z_1(k)| \leq l_2$.
In this case, along with the condition $\wp_1 + \wp_2 \leq 1$, inequality (7.15) can be further expressed as:

$$\Delta V(z) \leq -l_1|z_1(k)|^{\frac{1}{2}} - \wp_2|z_1(k)| \leq -l_1|z_1(k)|^{\frac{1}{2}}.$$

Since l_1 is chosen such that the condition $l_1 > \left(1 + \left|\frac{z_2(k)}{z_1(k)}\right|\right)^{\frac{1}{2}}$ holds, then

$$\Delta V(z) \leq -\left(V(z(k))\right)^{\frac{1}{2}} \tag{7.17}$$

Case III: If $\frac{\wp_1|z_1(k)|}{l_1} > |z_1(k)|^{\frac{1}{2}}$ and $\wp_2|z_1(k)| > l_2$.
Considering inequality (7.15) along with the constraint $\wp_1 + \wp_2 \leq 1$, we obtain:

$$\Delta V(z) \leq -l_1|z_1(k)|^{\frac{1}{2}} - l_2 \leq -\min(l_1|z_1(k)|^{\frac{1}{2}}, l_2).$$

Furthermore, given that $l_1 > \left(1 + \left|\frac{z_2(k)}{z_1(k)}\right|\right)^{\frac{1}{2}}$, it follows that

$$\Delta V(z) \leq -\min((V(z(k)))^{\frac{1}{2}}, l_2) \tag{7.18}$$

Case IV: If $\frac{\wp_1|z_1(k)|}{l_1} \leq |z_1(k)|^{\frac{1}{2}}$ and $\wp_2|z_1(k)| \leq l_2$.
In this case, inequality (7.15) simplifies to:

$$\Delta V(z) \leq -(V(z(k)) - |z_2(k)|).$$

Additionally, if $z_2(0) = 0$, then from (7.11), it directly follows that

$$\Delta V(z) \leq -V(z(k)) \tag{7.19}$$

Note from equation (7.19) that $\Delta V(z) \leq -V(z(k))$ implies $V(z(k+1)) \leq 0$ in a single sampling instant, indicating a dead-beat type response. Furthermore, in *Case IV*, if $z_2(0) \neq 0$, the evolution of (7.11) leads to the transition of *Case IV* into one of *Case I*, *Case II*, or *Case III*. Additionally, by considering the conditions derived in (7.16), (7.17), (7.18), and (7.19), we obtain the following:

$$\Delta V(z) \leq -\min(V(z(k)), (V(z(k)))^{\frac{1}{2}}, l_2) \tag{7.20}$$

Hence, the solution $z(k) \equiv 0$ is FTS. Furthermore, since $V(\cdot)$ is radially unbounded, and if $D = \mathbb{R}^2$, global FTS follows by similar arguments as presented above. This completes the proof. □

Remark 7.3 Note that under the conditions specified in *Case IV*, system (7.11) behaves like a linear system in the neighborhood of the origin. Moreover, if (7.19) holds, the system exhibits a dead-beat type response, converging to the origin in just one sample instant.

Remark 7.4 In (7.20), with the trajectory $V(z(k))$ entering the unit ball, it reaches the minimum of all the arguments, thereby stimulating a dead-beat type response, which results in convergence to the origin.

Remark 7.5 It may seem that when $z_1 \to 0, l_1 \to \infty$, but this is not the case because at this instant, $z_2 \to 0$ as well. Thus, l_1 will always satisfy the gain condition stated in the theorem.

7.3.2 Super-Twisting-Type Algorithm for Perturbed Discrete-Time Systems

Let us consider the following nonlinear discrete-time system.

$$\begin{aligned} z_1(k+1) =& \wp_1 z_1(k) - l_1 \mathrm{sign}(z_1(k)) \min\left(\frac{\wp_1|z_1(k)|}{l_1}, |z_1(k)|^{\frac{1}{2}}\right) + z_2(k), \\ z_2(k+1) =& \wp_2 z_1(k) - \mathrm{sign}(z_1(k)) \min\left(\wp_2|z_1(k)|, l_2\right) + \delta(k) \end{aligned} \tag{7.21}$$

where $z = [z_1, z_2]^\top \in D \subseteq \mathbb{R}^2$ denotes the system's states, $k \in \mathbb{Z}_{\geq 0}, \wp_1, \wp_2 \in (0, 1)$, and $l_1, l_2 \in \mathbb{R}_+$. The perturbation $\delta(k)$ is a bounded function such that $\delta : \mathbb{Z}_{\geq 0} \to \mathbb{R}$ and satisfies $|\delta(k)| \leq \delta_0$.

Again, we proceed with the analysis of the four cases as discussed for the unperturbed discrete-time system (7.11):

(i) If $\frac{\wp_1|z_1(k)|}{l_1} \leq |z_1(k)|^{\frac{1}{2}}$ and $\wp_2|z_1(k)| > l_2$,

(ii) If $\frac{\wp_1|z_1(k)|}{l_1} > |z_1(k)|^{\frac{1}{2}}$ and $\wp_2|z_1(k)| \leq l_2$,

(iii) If $\frac{\wp_1|z_1(k)|}{l_1} > |z_1(k)|^{\frac{1}{2}}$ and $\wp_2|z_1(k)| > l_2$,

(iv) If $\frac{\wp_1|z_1(k)|}{l_1} \leq |z_1(k)|^{\frac{1}{2}}$ and $\wp_2|z_1(k)| \leq l_2$.

We examine each of these cases and show that the solution $z(k)$ satisfies finite-time input-to-state stability.

We now present the following theorem to establish the finite-time input-to-state stability (FTISS) of the system given by (7.21).

Theorem 7.3 *Let us consider the system* (7.21) *with* $\wp_1 + \wp_2 \leq 1$, $l_1 > \left(1 + \left|\frac{z_2(k)}{z_1(k)}\right|\right)^{\frac{1}{2}}$, $z_1(k) \neq 0$, *and* $l_2 > \delta_0$. *Let there exists a continuous function* $V : \mathbb{Z}_{\geq 0} \times D \to \mathbb{R}_{\geq 0}$ *that satisfies:*

$$\alpha_1(\|z(k)\|) \leq V(k, z) \leq \alpha_2(\|z(k)\|), \forall k \in \mathbb{Z}_{\geq 0}, \forall z \in D \backslash \{0\}, \tag{7.22}$$

$$V(k, 0) = 0, \forall k \in \mathbb{Z}_{\geq 0} \text{ and} \tag{7.23}$$

$$\Delta V(k, z) \leq -\min(V(k, z), (V(k, z))^{\frac{1}{2}}, l_2) + \alpha_3(|\delta(k)|), \forall k \in \mathbb{Z}_{\geq 0}, \forall z \in D \backslash \{0\}, \tag{7.24}$$

where α_1, α_2, *and* $\alpha_3 \in$ *class* $\mathcal{K}$ *functions. Then, the system* (7.21) *is FTISS. Furthermore, if* $D = \mathbb{R}^2$, α_1 *and* $\alpha_2 \in$ *class* $\mathcal{K}_\infty$ *functions, and if* $\Delta V(k, z) \leq -\min\left(V(k, z), (V(k, z))^{\frac{1}{2}}, l_2\right) + \alpha_3(|\delta(k)|)$, *for all* $k \in \mathbb{Z}_{\geq 0}$ *and* $z \in \mathbb{R}^2 \setminus \{0\}$, *then the system* (7.21) *is globally FTISS.*

Proof: As discussed earlier, we consider the case where $z_1(k) \neq 0$. Let us define a candidate Lyapunov function as $V(z(k)) = |z_1(k)| + |z_2(k)|$. Then, the Lyapunov function at the next time step is given by $V(z(k+1)) = |z_1(k+1)| + |z_2(k+1)|$. The change in the Lyapunov energy is:

$$\begin{aligned}\Delta V(k,z) =& V(z(k+1)) - V(z(k)) \\ =& \left|\wp_1 z_1(k) - l_1 \text{sign}(z_1(k)) \min\left(\frac{\wp_1|z_1(k)|}{l_1}, |z_1(k)|^{\frac{1}{2}}\right) + z_2(k)\right| \\ &+ \left|\wp_2 z_1(k) - \text{sign}(z_1(k)) \min\left(\wp_2|z_1(k)|, l_2\right) + \delta(k)\right| \\ &- |z_1(k)| - |z_2(k)|\end{aligned}$$

Since $\wp_1|z_1(k)| \geq l_1 \min\left(\frac{\wp_1|z_1(k)|}{l_1}, |z_1(k)|^{\frac{1}{2}}\right)$ and $\wp_2|z_1(k)| \geq \min\left(\wp_2|z_1(k)|, l_2\right)$, we can further express:

$$\begin{aligned}\Delta V(k,z) \leq & -\left|l_1 \min\left(\frac{\wp_1|z_1(k)|}{l_1}, |z_1(k)|^{\frac{1}{2}}\right)\right| + |\delta(k)| \\ & - \left|\min\left(\wp_2|z_1(k)|, l_2\right)\right| + (\wp_1 + \wp_2 - 1)|z_1(k)|. \end{aligned} \tag{7.25}$$

From this point, different cases are analyzed, leading to the derivation of the difference inequality (7.25).

Case I: If $\frac{\wp_1|z_1(k)|}{l_1} \leq |z_1(k)|^{\frac{1}{2}}$ and $\wp_2|z_1(k)| > l_2$.
Given that $\wp_1 + \wp_2 \leq 1$, we can further write:

$$\Delta V(k,z) \leq -l_1 \min\left(\frac{\wp_1|z_1(k)|}{l_1}, |z_1(k)|^{\frac{1}{2}}\right) - \min\left(\wp_2|z_1(k)|, l_2\right) + |\delta(k)|.$$

It then follows that:

$$\begin{aligned}\Delta V(k,z) \leq & -\wp_1|z_1(k)| - l_2 + |\delta(k)| \\ \leq & -l_2 + |\delta(k)|. \end{aligned} \tag{7.26}$$

Case II: If $\frac{\wp_1|z_1(k)|}{l_1} > |z_1(k)|^{\frac{1}{2}}$ and $\wp_2|z_1(k)| \leq l_2$.
Given that $\wp_1 + \wp_2 \leq 1$, we can further express (7.25) as:

$$\begin{aligned}\Delta V(k,z) \leq & -l_1|z_1(k)|^{\frac{1}{2}} - \wp_2|z_1(k)| + |\delta(k)| \\ \leq & -l_1|z_1(k)|^{\frac{1}{2}} + |\delta(k)|. \end{aligned} \tag{7.27}$$

Moreover, if $l_1 > \left(1 + \left|\frac{z_2(k)}{z_1(k)}\right|\right)^{\frac{1}{2}}$, we obtain

$$\Delta V(k,z) \leq -\left(V(z(k))\right)^{\frac{1}{2}} + |\delta(k)|. \tag{7.28}$$

Case III: If $\frac{\wp_1|z_1(k)|}{l_1} > |z_1(k)|^{\frac{1}{2}}$ and $\wp_2|z_1(k)| > l_2$.
Considering (7.25) and the condition $\wp_1 + \wp_2 \leq 1$, we obtain:

$$\begin{aligned}\Delta V(k,z) \leq& -l_1|z_1(k)|^{\frac{1}{2}} - l_2 + |\delta(k)| \\ \leq& -\min\left(l_1|z_1(k)|^{\frac{1}{2}}, l_2\right) + |\delta(k)|. \end{aligned} \tag{7.29}$$

Furthermore, if $l_1 > \left(1 + \left|\frac{z_2(k)}{z_1(k)}\right|\right)^{\frac{1}{2}}$, we can write:

$$\Delta V(k,z) \leq -\min\left((V(z(k)))^{\frac{1}{2}}, l_2\right) + |\delta(k)|. \tag{7.30}$$

Case IV: If $\frac{\wp_1|z_1(k)|}{l_1} \leq |z_1(k)|^{\frac{1}{2}}$ and $\wp_2|z_1(k)| \leq l_2$.
In this case, (7.25) can be further expressed as:

$$\Delta V(k,z) \leq -(V(z(k)) - |z_2(k)|) + |\delta(k)|.$$

Additionally, if $z_2(0) = 0$, then from (7.21), it directly follows that:

$$\Delta V(k,z) \leq -V(z(k)) + |\delta(k)|. \tag{7.31}$$

Moreover, in *Case IV*, if $z_2(0) \neq 0$, the time evolution of (7.21) implies that *Case IV* transitions into either *Case I*, *Case II*, or *Case III*. By considering the conditions derived in (7.26), (7.28), (7.30), and (7.31), we conclude:

$$\Delta V(k,z) \leq -\min\left(V(z(k)), (V(z(k)))^{\frac{1}{2}}, l_2\right) + \alpha_3(|\delta(k)|). \tag{7.32}$$

Thus, the system (7.21) is finite-time input-to-state stable. Moreover, since $V(\cdot)$ is radially unbounded, and assuming $D = \mathbb{R}^2$, global FTISS follows from the arguments made earlier. This concludes the proof. □

7.3.3 Simulation Results for the Super-Twisting-like Algorithm

The MATLAB simulation was conducted using three distinct sets of initial conditions to assess the performance and effectiveness of the proposed algorithm. To ensure a comprehensive evaluation, the system parameters were selected as $l_1 = 3$, $l_2 = 5$, $\wp_1 = 0.6$, and $\wp_2 = 0.4$.

The trajectories of the system states without any perturbation are depicted in Fig. 7.3. Specifically, the trajectory of $z_1(k)$ is shown in Fig. 7.3a, while the trajectory of $z_2(k)$ is presented in Fig. 7.3b. From Fig. 7.3, it is evident that the super-twisting-like

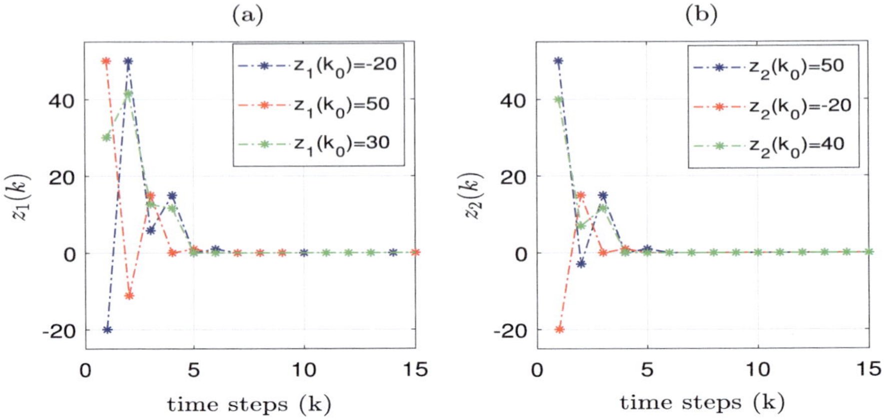

Fig. 7.3 State evolution of the unperturbed DTS

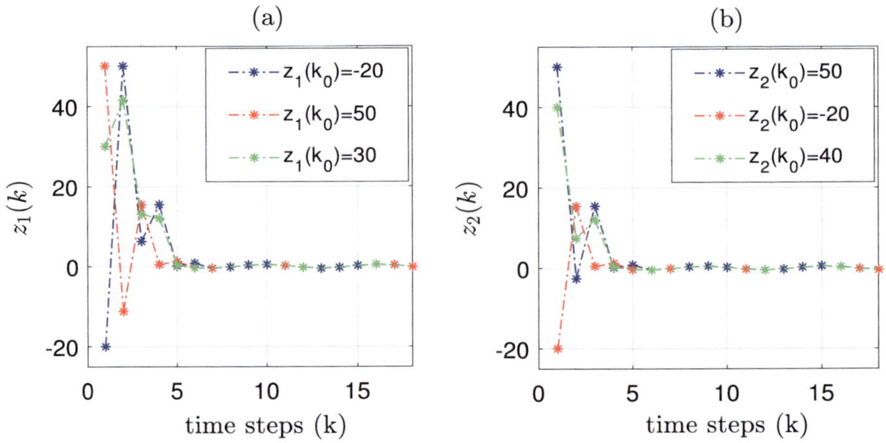

Fig. 7.4 State evolution of the perturbed DTS

algorithm effectively drives the system states to converge to the origin in finite-time steps.

The corresponding phase portrait of the system, shown in Fig. 7.5a, further illustrates that the algorithm ensures the states approach zero within a finite time frame.

Next, when the system is subjected to a Model-Triggered Bounded Perturbation (MTBP) of the form 0.5 $\sin(k)$, the resulting trajectories are presented in Fig. 7.4. As observed in Fig. 7.4a, b, the system states remain confined within the vicinity of the origin when MTBP is applied. This behavior is also evident in the phase portrait shown in Fig. 7.5b, which confirms that the states stay close to the origin despite the perturbation.

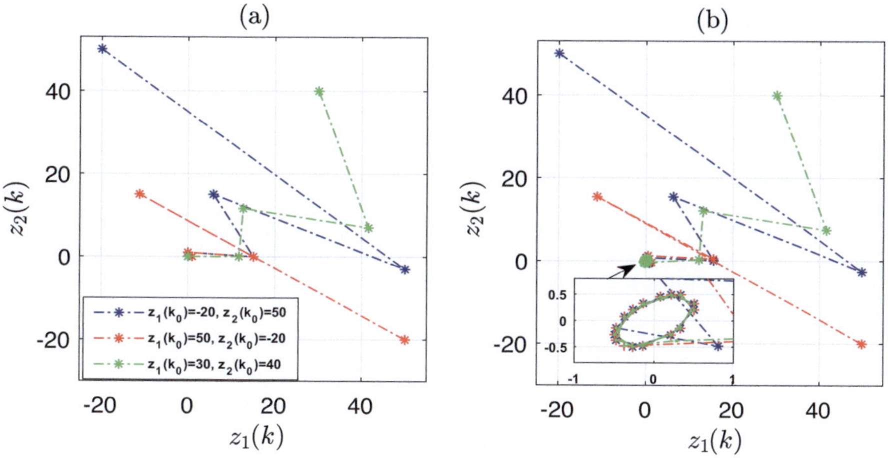

Fig. 7.5 Phase portrait: **a** Unperturbed DTS and **b** Perturbed DTS

Remark 7.6 The algorithm proposed for the unperturbed DTS in (7.11) drives the system states to the origin, rather than merely keeping them in the vicinity of the origin. In contrast, most Euler-discretization-based Super-Twisting Algorithms (STAs) ([4, 7, 8]) result in the system states eventually residing in an invariant set that contains the origin, but they do not guarantee convergence to the origin.

7.4 Difference Equation with Minima Based STO

This section discusses the development of a new type of observer, which is derived from the proposed algorithm. This observer is designed to estimate the system states in finite time.

7.4.1 STO Construction

Consider the following dynamical system:

$$
\begin{aligned}
z_1(k+1) =& z_2(k) \\
z_2(k+1) =& f(z_1(k), z_2(k), u(k)) + \delta(k)
\end{aligned}
\tag{7.33}
$$

The system output is given by $y(k) = z_1(k)$, where $z = [z_1, z_2]^\top \in D \subseteq \mathbb{R}^2$ represents the system states, $k \in \mathbb{Z}_{\geq 0}$ is the discrete-time index, $u \in \mathbb{R}$ is the input, and $\delta(k)$ denotes the Model-Triggered Bounded Perturbation (MTBP), satisfying $|\delta(k)| \leq \delta_0$.

We then derive the difference equation with a minima-based Super-Twisting Observer (STO) as follows:

$$
\begin{aligned}
\hat{z}_1(k+1) &= \wp_1(\hat{z}_1(k) - z_1(k)) - l_1 \text{sign}(\hat{z}_1(k) - z_1(k)) \\
&\quad \min\left(\frac{\wp_1|\hat{z}_1(k) - z_1(k)|}{l_1}, |\hat{z}_1(k) - z_1(k)|^{\frac{1}{2}}\right) + \hat{z}_2(k), \\
\hat{z}_2(k+1) &= \wp_2(\hat{z}_1(k) - z_1(k)) - \text{sign}(\hat{z}_1(k) - z_1(k)) \\
&\quad \min(\wp_2|(\hat{z}_1(k) - z_1(k))|, l_2) + \hat{f}(z_1(k), \hat{z}_2(k), u(k))
\end{aligned}
\tag{7.34}
$$

where $l_1, l_2 \in \mathbb{R}_+$ and $\wp_1, \wp_2 \in (0, 1)$ are positive constants chosen appropriately. We define the error variables as $e_1 := \hat{z}_1(k) - z_1(k)$ and $e_2 := \hat{z}_2(k) - z_2(k)$. The error dynamics take the form of the super-twisting-like algorithm discussed in (7.11):

$$
\begin{aligned}
e_1(k+1) =& \wp_1 e_1(k) - l_1 \text{sign}(e_1(k)) \min\left(\frac{\wp_1|e_1(k)|}{l_1}, |e_1(k)|^{\frac{1}{2}}\right) + e_2(k), \\
e_2(k+1) =& \wp_2 e_1(k) - \text{sign}(e_1(k)) \min(\wp_2|e_1(k)|, l_2) + \tilde{f}(z_1(k), \hat{z}_2(k), u(k)) - \delta(k)
\end{aligned}
\tag{7.35}
$$

where $\tilde{f}(z_1(k), \hat{z}_2(k), u(k)) := \hat{f}(z_1(k), \hat{z}_2(k), u(k)) - f(z_1(k), z_2(k), u(k))$. Furthermore, we can express this as:

$$
\begin{aligned}
e_1(k+1) =& \wp_1 e_1(k) - l_1 \text{sign}(e_1(k)) \min\left(\frac{\wp_1|e_1(k)|}{l_1}, |e_1(k)|^{\frac{1}{2}}\right) + e_2(k), \\
e_2(k+1) =& \wp_2 e_1(k) - \text{sign}(e_1(k)) \min(\wp_2|e_1(k)|, l_2) + \phi(k)
\end{aligned}
\tag{7.36}
$$

where $\phi(k) := \tilde{f}(z_1(k), \hat{z}_2(k), u(k)) - \delta(k)$, and it satisfies $|\phi(k)| \le \phi_0$.

The following corollary follows directly from Theorem 6.8.

Corollary 7.1 *Let us consider the dynamics in* (7.36) *with* $\wp_1 + \wp_2 \le 1$, $l_1 > \left(1 + \left|\frac{e_2(k)}{e_1(k)}\right|\right)^{\frac{1}{2}}$, $e_1(k) \neq 0$, *and* $l_2 > \phi_0$. *Let there exists a continuous function* $V : \mathbb{Z}_{\ge 0} \times D \to \mathbb{R}_{\ge 0}$ *that satisfies:*

$$\alpha_4(\|e\|) \le V(k, e) \le \alpha_5(\|e\|), \forall k \in \mathbb{Z}_{\ge 0}, \forall e \in D \backslash \{0\}, \tag{7.37}$$

$$V(k, 0) = 0, \forall k \in \mathbb{Z}_{\ge 0} \text{ and} \tag{7.38}$$

$$\Delta V(k, e) \le -\min(V(k, e), (V(k, e))^{\frac{1}{2}}, l_2) + \alpha_6(|\phi(k)|), \forall k \in \mathbb{Z}_{\ge 0}, \forall e \in D \backslash \{0\} \tag{7.39}$$

where $\alpha_4, \alpha_5, \alpha_6 \in$ *class* $\mathcal{K}$ *functions. Then, the system in* (7.36) *is finite-time input-to-state stable. Furthermore, if* $D = \mathbb{R}^2$, *with* α_4 *and* $\alpha_5 \in$ *class* $\mathcal{K}_\infty$ *functions, and if* $\Delta V(k, e) \le -\min\left(V(k, e), (V(k, e))^{\frac{1}{2}}, l_2\right) + \alpha_6(|\phi(k)|)$, *for all* $k \in \mathbb{Z}_{\ge 0}$ *and* $e \in \mathbb{R}^2 \setminus \{0\}$, *then the system in* (7.36) *is globally finite-time input-to-state stable.*

Proof: We choose a similar Lyapunov candidate given by $V(e(k)) = |e_1(k)| + |e_2(k)|$. Consequently, $V(e(k+1)) = |e_1(k+1)| + |e_2(k+1)|$. The variation in the

Lyapunov energy is then expressed as:

$$\begin{aligned}\Delta V(k,e) =& \Big|\wp_1 e_1(k) - l_1 \text{sign}(e_1(k)) \min\left(\frac{\wp_1|e_1(k)|}{l_1}, |e_1(k)|^{\frac{1}{2}}\right) \\ &+ e_2(k)\Big| + \Big|\wp_2 e_1(k) - \text{sign}(e_1(k)) \min(\wp_2|e_1(k)|, l_2) - \phi(k)\Big| \\ &- |e_1(k)| - |e_2(k)|\end{aligned}$$

Furthermore, by applying a similar analysis to that used in deriving (7.26), (7.28), (7.30), and (7.31) for the super-twisting-like algorithm applied to the perturbed DTS, we can further express:

$$\Delta V(k,e) \leq -\min(V(k,e), (V(k,e))^{\frac{1}{2}}, l_2) + \alpha_6(|\phi(k)|) \tag{7.40}$$

Thus, the error dynamical system in (7.36) demonstrates FTISS. Moreover, since $V(\cdot)$ is radially unbounded, and assuming $D = \mathbb{R}^2$, global FTISS follows from the similar arguments presented in the previous section. This concludes the proof. □

7.4.2 STO Development for Pendulum System

In this section, we address the state observation problem for a perturbed pendulum system. An STO is designed for this system, and we demonstrate that the state estimation is achieved in finite time. Consider the following dynamical system:

$$\begin{aligned}z_1(k+1) &= z_1(k) + z_2(k) \\ z_2(k+1) &= z_2(k) - \frac{mgl}{2J}\sin(z_1(k)) - \frac{B}{J}z_2(k) + \frac{1}{J}u(k) + \delta(k)\end{aligned} \tag{7.41}$$

The system output is defined as $y(k) = z_1(k)$, where $z_1(k)$ represents the pendulum angle from the mean position and $z_2(k)$ is the angular velocity. The system parameters include the mass of the pendulum m, the acceleration due to gravity g, the length of the pendulum l, the inertia of the pendulum arm J, the friction coefficient B, and the input $u(k)$, which is given by $u(k) = 0.5\sin(0.6k)$. The bounded perturbation is modeled as $\delta(k) = 0.2\sin(0.3k)$.

The design of the difference equation with the minima-based STO for the pendulum system follows the form described in (7.34). The error dynamics of the observer are governed by the equations in (7.35), ensuring that the observer achieves finite-time input-to-state stability (FTISS).

For the simulation, the following parameter values are used: the mass $m = 1.1$ kg, the length $l = 1$ m, the gravitational acceleration $g = 9.81\ \frac{\text{m}}{\text{s}^2}$, and the friction coefficient $B = 0.18\ \frac{\text{kg}\cdot\text{m}}{\text{s}^2}$. The initial conditions for the system are set as $z_1(0) = 5$ and $z_2(0) = 2$, while the initial estimates for the observer are $\hat{z}_1(0) = 0$ and $\hat{z}_2(0) = 0$.

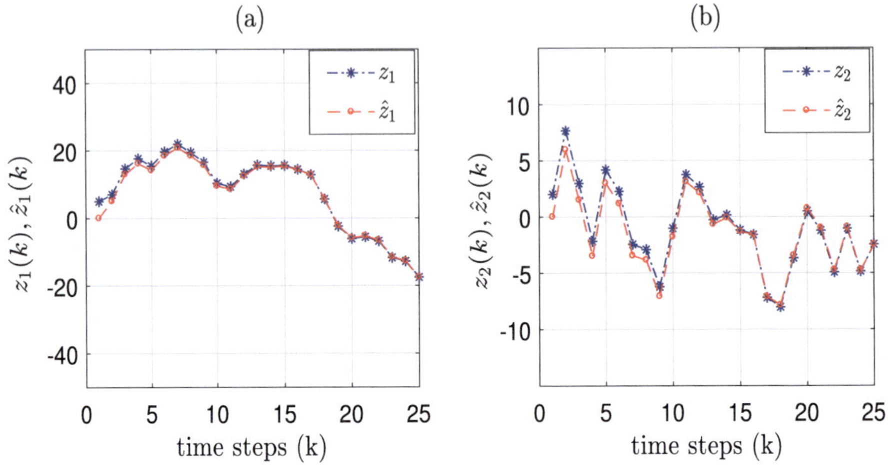

Fig. 7.6 Evolution of the system states and the super-twisting observer

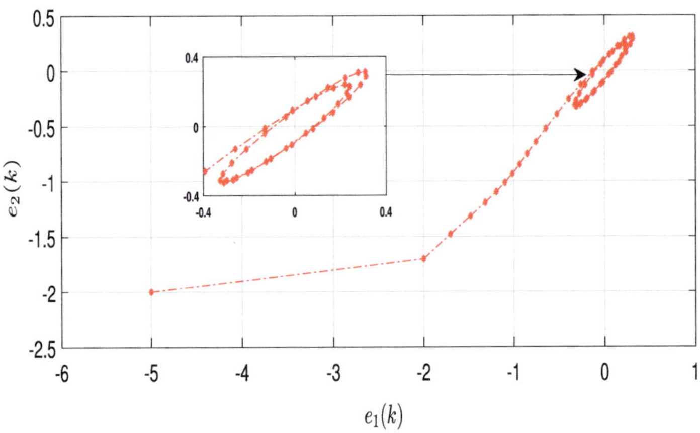

Fig. 7.7 Phase portrait of the error variables of the system

The super-twisting observer parameters are chosen as $l_1 = 4$, $l_2 = 6$, $\wp_1 = 0.5$, and $\wp_2 = 0.3$.

The trajectories of both the system states and the observer states are depicted in Fig. 7.6. From the figure, it is evident that the state estimates provided by the super-twisting observer (STO) closely follow the actual states of the perturbed system, with the estimation process completing in finite-time steps.

Additionally, the phase portrait of the system error variables is presented in Fig. 7.7. The plot illustrates that the proposed algorithm effectively drives the error dynamics of the system towards an ultimately bounded region that contains the origin, and this convergence occurs within a finite number of steps.

7.5 Conclusion

This chapter explored the DEM-based discrete-time twisting-like algorithm and discrete-time super-twisting-like algorithm, offering an alternative to traditional Euler-discretization-based methods. Both algorithms were thoroughly examined in the context of unperturbed and perturbed systems, with detailed proofs demonstrating finite-time stability and finite-time input-to-state stability. Furthermore, the chapter highlighted a significant application of the discrete-time super-twisting-like algorithm in the design of a discrete-time super-twisting observer. An illustrative example involving a pendulum system was presented, incorporating both known and unknown inputs for the observer. Simulation results clearly showcased the effectiveness of the DEM-based discrete-time super-twisting-like algorithm.

The next chapter shifts focus to the practical implementation of the difference equation with minima-based discrete sliding mode control.

References

1. Prasun, P., Kamal, S., Pandey, S., Bartoszewicz, A., Ghosh, S.: A minimum operator based discrete-time sliding mode control. IEEE Trans. Autom. Control **69**(11), 7871–7876 (2024)
2. Prasun, P., Pandey, S., Kamal, S., Ghosh, S., Xiong, X.: A minimum operator based discrete-time super-twisting-like algorithm. IEEE Trans. Circ. Syst. II: Expr. Briefs **71**(1), 286–290 (2023)
3. Koch, S., Reichhartinger, M.: Discrete-time equivalents of the super-twisting algorithm. Automatica **107**, 190–199 (2019)
4. Salgado, I., Kamal, S., Bandyopadhyay, B., Chairez, I., Fridman, L.: Control of discrete time systems based on recurrent super-twisting-like algorithm. ISA Trans. **64**, 47–55 (2016)
5. Xiong, X., Kikuuwe, R., Kamal, S., Jin, S.: Implicit-Euler implementation of super-twisting observer and twisting controller for second-order systems. IEEE Trans. Circ. Syst. II: Express Briefs **67**(11), 2607–2611 (2020)
6. Xiong, X., Bai, Y., Shi, R., Kamal, S., Wang, Y., Lou, Y.: Discrete-Time twisting algorithm implementation with Implicit-Euler ZOH discretization method. IEEE Trans. Circ. Syst. II: Express Briefs **69**(8), 3435–3439 (2022)
7. Yan, Y., Yu, X., Sun, C.: Discretization behaviors of a super-twisting algorithm based sliding mode control system. Int. Works. Recent Adva. Sliding Modes (RASM) 1–5 (2015)
8. Salgado, I., Chairez, I., Bandyopadhyay, B., Fridman, L., Camacho, O.: Discrete-time non-linear state observer based on a super twisting-like algorithm. IET Contr. Theo. Appl. **8**, 803–812 (2014)

Chapter 8
Application with Simulation and Experimental Validation

Abstract This chapter focuses on the practical implementation and real-world application of the theoretical concepts developed earlier. It bridges the gap between theory and practice by comparing simulation results with actual experimental outcomes. The primary objective is to validate the proposed discrete-time sliding mode control theory, specifically the difference equation with minima-based approach, through experimental testing. By analyzing the performance of the proposed control strategy in real-world scenarios, the chapter demonstrates its effectiveness and robustness in comparison to the theoretical predictions from the simulations. Additionally, the effect of sampling time is analyzed, revealing that shorter sampling intervals result in increased control effort, as demonstrated through simulation.

8.1 Application to a Coupled Tank System

In this section, both MATLAB/Simulink simulations and experimental tests are presented to verify the effectiveness and accuracy of the proposed reaching law for discrete-time sliding mode control. While the simulations provide a controlled environment to test the theoretical model, the experimental component is crucial for validating the practical applicability of the reaching law in real-world conditions. This combination of simulation and experimentation allows for a comprehensive evaluation of the proposed method, particularly in assessing how well the theoretical results hold when subjected to the complexities of real-world dynamics and uncertainties. The experimental part is particularly critical for investigating key performance metrics, such as the successful mitigation of chattering and the precision of reference tracking. These aspects are vital for demonstrating that the proposed reaching law not only works in theory but also performs effectively in practical applications where real-world challenges, like disturbances and model inaccuracies, can influence system behavior.

The coupled tank system is used as the test platform, with the schematic diagram shown in Fig. 8.1 illustrating the model. The goal is to control the water level in tank 1 to match the desired set-point. In its present setup (Configuration 1), the control input is applied directly to the inlet of tank 1.

S. Kamal and P. Prasun, *A Minimum Function-Based Discrete Sliding Mode Control: Theory and Applications*, Studies in Systems, Decision and Control 621,
https://doi.org/10.1007/978-981-95-1161-7_8

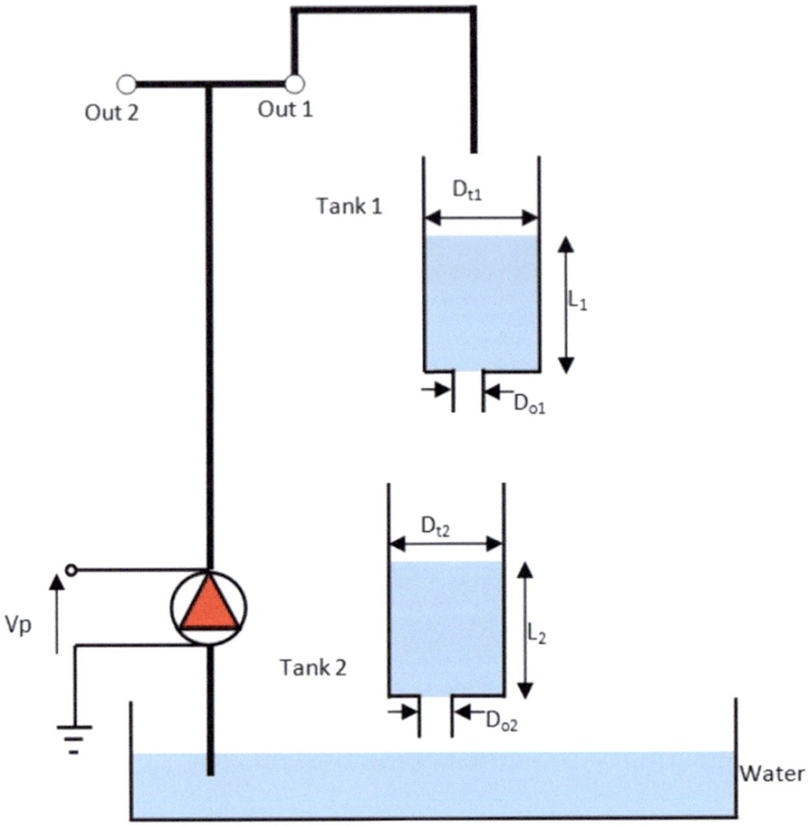

Fig. 8.1 Block diagram representation of the coupled water tank system

The mathematical model for this configuration is given by:

$$\frac{d}{dt}L_1 = \frac{1}{A_{t1}}\left[K_p V_p - A_{o1}\sqrt{2gL_1}\right] \tag{8.1}$$

In the equation, L_1 represents the water level in tank 1, V_p denotes the pump input voltage, K_p is the proportionality constant, A_{t1} and A_{o1} refer to the cross-sectional area of tank 1 and its outlet hole, respectively, and g is the acceleration due to gravity.

We define the state variable as $z(t) = L_1$, and the control input variable as $u(t) = V_p$.

Thus, we can transform the system variables and express the system dynamics as follows:

$$\dot{z}(t) = \frac{1}{A_{t1}}\left[K_p u(t) - A_{o1}\sqrt{2gz(t)}\right] \tag{8.2}$$

The parameters of the tank system are as follows: $A_{o1} = 0.1781\,\text{cm}^2$, $A_{t1} = 15.517\,\text{cm}^2$, $K_p = 3.3\,\text{cm}^3/\text{s/V}$, and $g = 981\,\text{cm/s}^2$. The maximum voltage that can be applied to the tank system is 22 V. We linearize the system around the operating point z_{1o}, leading to the following equations:

$$\dot{z}(t) = -\frac{A_{o1}}{A_{t1}}\sqrt{\frac{g}{2z_{1o}}}z(t) + \frac{K_p}{A_{t1}}u(t) \tag{8.3}$$

Additionally, we discretize the system with a sampling time T_s to formulate the proposed reaching law-based discrete sliding mode control. The discrete-time model for the tank system is derived using zero-order hold on the input, with a sampling time of $T_s = 0.1$ sec and an operating point $z_{1o} = 15$ cm, as follows:

$$z(k+1) = 0.9935z(k) + 0.0212u(k) \tag{8.4}$$

Let the desired water level trajectory for tank 1 be denoted as $z_d(k)$. The tracking error is defined as $e(k) = z_d(k) - z(k)$. The dynamics of the error coordinates are critical for assessing the performance of the control system. In the subsequent section, we will analyze these dynamics to determine the control signal required to maintain the desired water level trajectory. The error dynamics are expressed as follows:

$$\begin{aligned} e(k+1) &= z_d(k+1) - z(k+1) \\ &= z_d(k+1) - \{0.9935z(k) + 0.0212u(k)\} \\ &= z_d(k+1) - \{0.9935(z_d(k) - e(k)) + 0.0212u(k)\} \end{aligned} \tag{8.5}$$

Furthermore, the control law is computed based on the proposed reaching law as follows:

$$\begin{aligned} u(k) = &-\frac{1}{0.0212}\Bigg\{ - z_d(k+1) + 0.9935(z_d(k) - e(k)) + e(k) \\ &- T_s\,\text{sign}(e(k))\min\left(\frac{|(e(k))|}{T_s}, \gamma\right)\Bigg\} \end{aligned} \tag{8.6}$$

8.2 Results and Discussions

In this section, we aim to demonstrate the effectiveness of the proposed reaching law through both simulation results and experimental validation. The Quanser setup for the coupled tank system is shown in Fig. 8.2. For this system, we select the value of the reaching law design parameter as $\gamma = 2$.

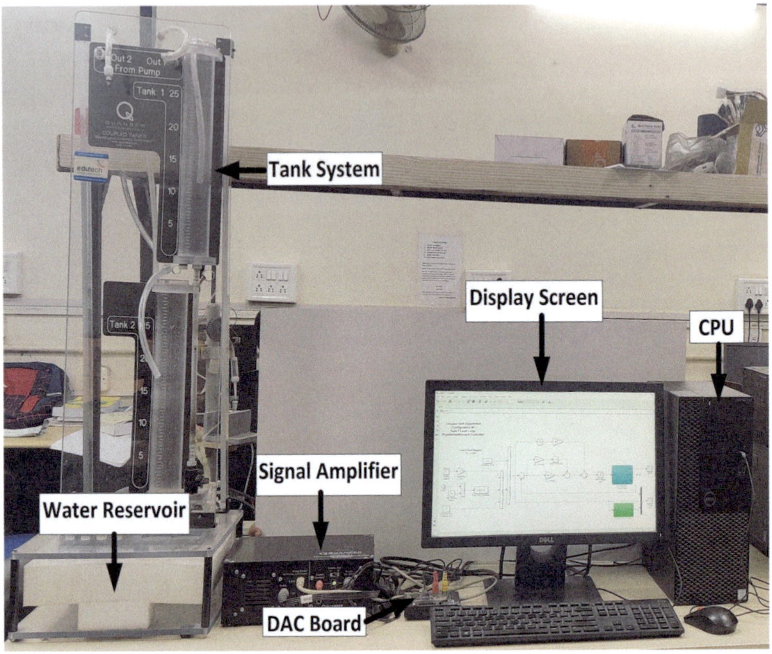

Fig. 8.2 Experimental configuration of the coupled water tank system

8.2.1 Unperturbed Case

The simulation results for the unperturbed case are presented in Fig. 8.3. The reference tracking of the water level is shown in Fig. 8.3a, where it is evident that the system state tracks the reference trajectory perfectly, without any oscillations. Additionally, from the control input shown in Fig. 8.3b, it can be observed that there is no chattering present in the control signal.

The corresponding experimental results for the undisturbed case are shown in Fig. 8.4. The water level tracking is illustrated in Fig. 8.4a, and the control input is depicted in Fig. 8.4b. The experimental results closely match the simulation outcomes, particularly in terms of the water level tracking. However, it is important to note that the control input exhibits rapid changes when the reference value is held constant. This behavior is likely due to the inherent uncertainties arising from sensor noise and other practical factors.

Fig. 8.3 Simulation results for the unperturbed system: **a** Water level tracking response, **b** Control input applied

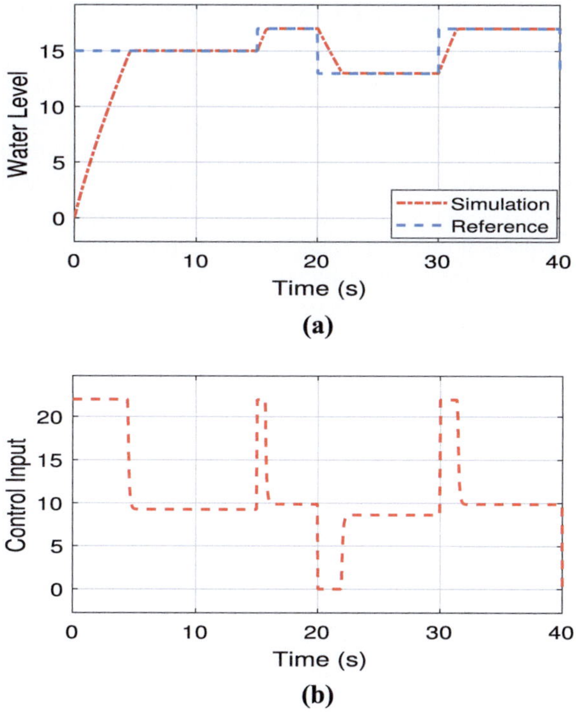

Fig. 8.4 Experimental results for the unperturbed system: **a** Water level tracking response, **b** Applied control input

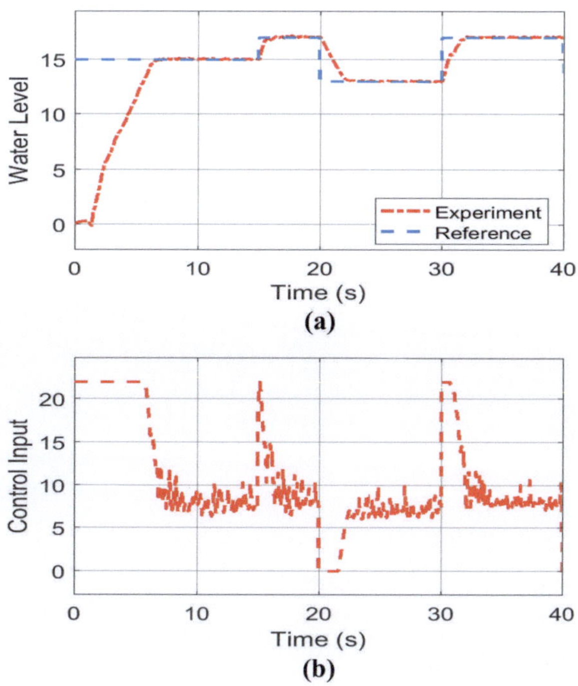

8.2.2 Perturbed Case

The simulation results for the perturbed case are shown in Fig. 8.5. In this simulation, a sinusoidal perturbation is introduced into the tank dynamics. Despite the presence of such perturbations, the system successfully tracks the reference water level trajectory without any oscillations, as seen in Fig. 8.5a. Furthermore, the applied control input, displayed in Fig. 8.5b, demonstrates no chattering. However, small amplitude ripples are present due to the added perturbation.

The corresponding experimental results for the perturbed system are presented in Fig. 8.6. In these experiments, perturbations are introduced by intermittently opening and closing a thin outlet at the bottom of the tank. The water level tracking response is shown in Fig. 8.6a, and the corresponding control input is depicted in Fig. 8.6b. The water level tracking closely matches the simulation results. However, it is worth noting that the control input exhibits rapid changes when the reference value is held constant, which may be attributed to the external perturbations and inherent uncertainties caused by sensor noise.

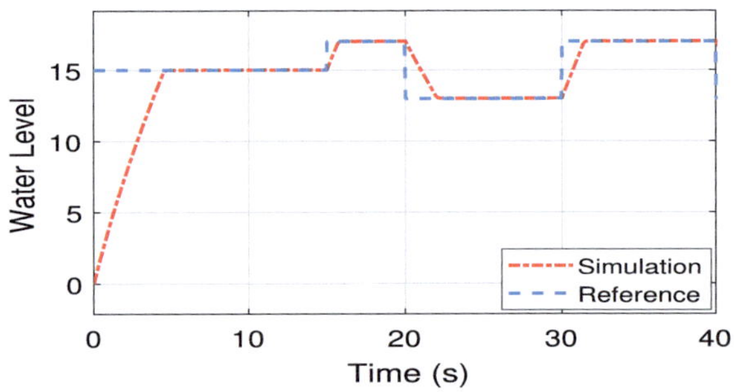

(a)

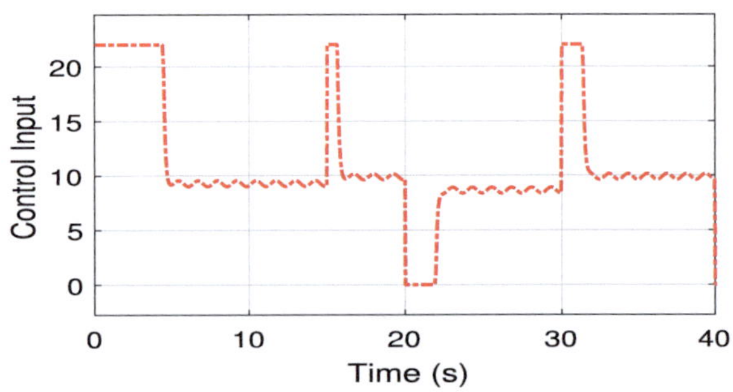

(b)

Fig. 8.5 Simulation results for the perturbed system: **a** Reference water level tracking, **b** Control input applied

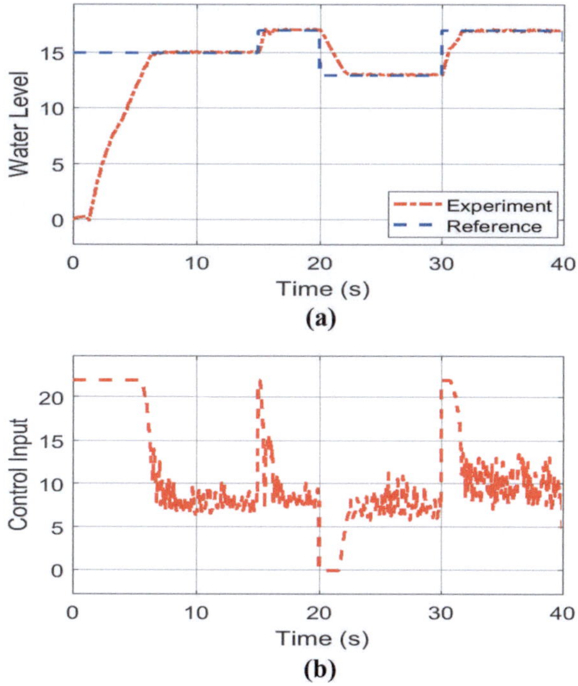

Fig. 8.6 Experimental results for the perturbed system: **a** Water level tracking response, **b** Applied control input

8.3 Effect of Sampling Time

In this section, we consider a simple first-order discrete-time system to show the dependency on the sampling time. The lesser the sampling time, more is the control effort and vice-versa. To visualize this, let us consider the following system:

$$z(k+1) = z(k) + T_s u(k) \tag{8.7}$$

where $z \in D \subset \mathbb{R}$ is the system state, $u \in \mathbb{R}$ is the control input, and $T_s \in \mathbb{R}_+$ is the sampling time. We design the control law as

$$u(k) = \frac{1}{T_s}\{-\text{sign}(z(k)) \min\{|z(k)|, \gamma\}\} \tag{8.8}$$

where $\gamma \in \mathbb{R}_+$.

To visualize the effect of sampling time, we simulate the above system with three different values of sampling time keeping initial condition and design constant fixed. We considered $z(0) = 50$, $\gamma = 3$, and $T_s = 0.9, 0.5, 0.3$.

From Fig. 8.7, we can see that as the sampling time is reduced, the convergence rate of system state increases, however chattering phenomenon also becomes prominent. Similarly, in Fig. 8.8, one can notice the increase in the amplitude of the control signal

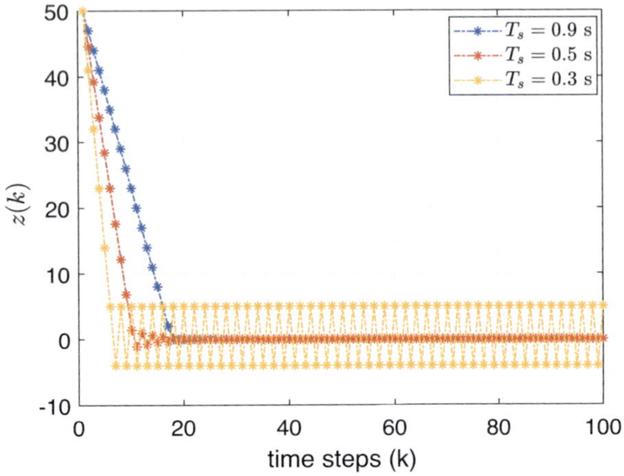

Fig. 8.7 Evolution of system state for different sampling time

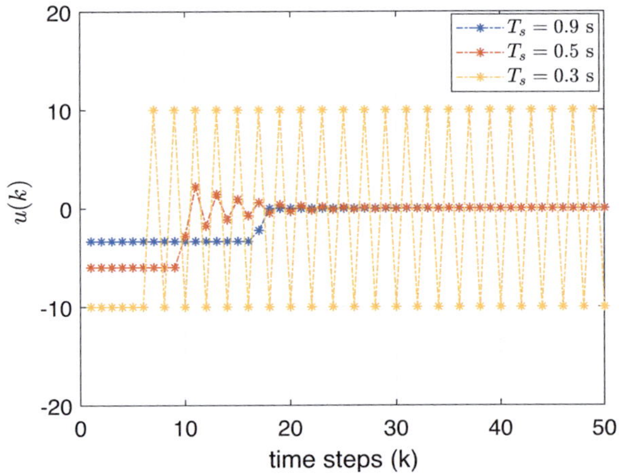

Fig. 8.8 Control effort for different sampling time

as the sampling time is reduced. From the design perspective, one should arrive at a suitable trade-off between speed of convergence and the amount of control effort required.

8.4 Conclusion

In this chapter, we examined the practical implementation of the proposed reaching law based on a difference equation with minima for designing discrete-time sliding mode control in a coupled tank system. Specifically, we applied this approach to a Quanser coupled tank system operating in Configuration 1. Two scenarios were considered: one without any external perturbation and another with the presence of an external perturbation. For both cases, we presented both simulation and experimental results.

The experimental water level tracking in both scenarios demonstrated excellent performance, closely matching the expected results from the simulation. In terms of control effort, the unperturbed system showed rapid variations in the control signal, which were attributed to sensor noise. On the other hand, in the perturbed system, rapid changes in the control signal were observed, caused by a combination of external disturbances and sensor noise. Moreover, the influence of sampling time was investigated, showing that shorter sampling intervals lead to greater control effort, as evidenced by simulation results.

In conclusion, the proposed controller successfully ensured precise water level tracking within finite time. The experimental results were consistent with the simulation results, validating the effectiveness of the proposed approach in real-world applications.

Chapter 9
Conclusions and Future Perspectives

Abstract In this chapter, we bring closure to the comprehensive body of work presented in this monograph, emphasizing its notable contributions while acknowledging its inherent limitations. Additionally, we identify prospective avenues for further exploration and investigation.

9.1 Conclusions

The main objective of this monograph was to develop and apply the theory of sliding mode control based on difference equations with minima, specifically tailored for discrete-time systems. Through this work, we sought to tackle ongoing challenges in the advancement of discrete sliding mode methods. Each chapter of this monograph makes distinctive contributions, and we summarize the main findings below:

Chapter 3 commenced with a comprehensive overview of the historical evolution and theoretical underpinnings of discrete-time sliding mode control, rooted in the foundational concepts of continuous-time sliding mode control. We initially outlined a basic design framework for CSMC, then traced the progression of DSMC, highlighting the key distinctions in their respective design methodologies. Furthermore, a pivotal classification was introduced to distinguish between switching and non-switching reaching law formulations within these control paradigms. These foundational discussions provided the necessary context and theoretical basis for the research contributions presented by the authors.

In the exploration detailed in Chap. 4, we introduced two reaching laws specifically designed for discrete-time systems, leveraging difference equation with minima. These reaching laws not only effectively eliminate chattering, ensuring smoother operation, but also contribute to improved system robustness. This improvement is achieved by strategically reducing the size of the quasi-sliding mode band width, without putting too much stress on the control effort. The delicate balance struck by these reaching laws showcases their efficacy in addressing nuanced challenges within discrete-time systems. Theoretical analysis and extensive simulation studies were

S. Kamal and P. Prasun, *A Minimum Function-Based Discrete Sliding Mode Control: Theory and Applications*, Studies in Systems, Decision and Control 621,
https://doi.org/10.1007/978-981-95-1161-7_9

conducted to validate the effectiveness and robustness of the proposed methodology in achieving slidng mode for unperturbed case and quasi-sliding mode for perturbed case.

Chapter 5 expanded upon the reaching laws based on the difference equation with minima, introducing a shift to rate-regulatory function-based reaching laws. The primary aim was to harness advantages related to the variable rate of change of the sliding variable, facilitated by the proportional term. This modification resulted in a significantly accelerated rate of change of the sliding variable. Notably, we demonstrated that this rate of change is effectively upper-bounded by the design parameters, offering an additional advantageous feature. We conducted the simulation on numerical example as well as on pendulum system. For pendulum system we conducted a comparative analysis among various methods and expectedly the findings revealed that the proposed method outperformed others.

Chapter 6 of the monograph set out to explore the Lyapunov characterization of discrete sliding mode based on difference equation with minima. In pursuit of this objective, we established the Lyapunov conditions for the finite-time input-to-state stability of perturbed discrete-time systems, offering insights into the settling time function. Moreover, we conducted a comparative study through simulations, highlighting the superior performance of the proposed laws in comparison to existing ones. An intriguing observation arising from this analysis is the attainment of the least ultimate bound for the sliding variable trajectory with satisfactory transient behavior.

Chapter 7 delved into the development of twisting-like and super-twisting-like algorithms, grounded in the framework of a difference equation with minima. Unlike conventional discrete-time algorithms which are often Euler-discretized, our proposed designs showcased distinctiveness. By implementing these designs, we successfully ensured the convergence of the system states to an equilibrium point for unperturbed systems and to an invariant set for perturbed systems, all within finite time. Furthermore, we leveraged the discrete super-twisting-like algorithm to design a super-twisting observer, extending its applicability. This observer, in turn, robustly estimated unknown state variables for a perturbed pendulum system within a finite timeframe.

Chapter 8 centers on the application of the suggested reaching law based on a difference equation with minima, and delves into the intricacies of the applied reaching law, providing a comprehensive analysis of its performance in the context of the couple tank system. The results not only demonstrate the robustness of the proposed approach but also highlight its adaptability to real-world experimental conditions. The proposed law was successfully implemented in the couple tank system, and we showcased the outcomes through simulation and experimentation. The level control of the tank system was conducted under unperturbed and perturbed conditions. In both scenarios, the successful attainment of the desired water level underscores the efficacy of the proposed laws. The study also analyzed the role of sampling time, demonstrating through simulations that shorter intervals are associated with increased control effort.

9.2 Future Perspectives

Although this monograph has significantly contributed to the field of discrete-time sliding mode control, it is crucial to recognize that the proposed design comes with certain limitations, suggesting possibilities for further progress. As a result, the subsequent sections outline these challenges and pinpoint potential avenues for future research and development:

- Exploring higher-order discrete sliding mode control stands out as a promising avenue that could significantly diminish the size of the quasi-sliding mode band and enhance the robust characteristics of sliding mode control. The implementation of higher-order discrete sliding mode involves elevating the relative degree of the sliding variable. Investigating the application of a difference equation with minima-based discrete sliding mode control, incorporating reaching laws with higher relative degrees, could prove to be an intriguing exploration. Notably, disturbances introduced in the reaching law from the rth (for r^{th}-order sliding mode) time instants ago lead to the confinement of system states within a much narrower band.
- Exploring more intricate and practical system models is imperative, particularly in addressing challenges inherent in real-world scenarios. This includes investigating stochastic systems to encompass a broad spectrum of uncertain dynamical systems. The nonlinearities and uncertainties in such systems may manifest in a probabilistic manner with specific types and intensities. It becomes essential to analyze these aspects separately, taking into account the stochastic behavior of the system, in order to effectively mitigate the adverse effects resulting from these uncertainties.
- Enhancing robustness properties by narrowing the quasi-sliding mode band width is an extensively researched area and is expected to remain a focus in the years to come. The pursuit of refining proximity to the sliding manifold continues to be a significant aspect. Despite the loss of the invariance property in discrete sliding mode control, there remains room for improvement in the degree of robustness, making it a pertinent area for further exploration.
- Examining robustness properties concerning more generalized uncertainties presents an intriguing avenue for research. The uncertainties affecting system performance can fall into either matched or unmatched categories. Throughout this monograph, the assumption has been that the system is subject to matched-type bounded uncertainty, and the analysis has been specifically tailored to this case. However, the stability properties of the closed-loop system may differ, or even worse, the system could become unstable under unmatched uncertainties. Therefore, a more comprehensive analysis in this direction could provide a better understanding of the limitations inherent in the research conducted in this monograph.

MIX
Papier aus verantwortungsvollen Quellen
Paper from responsible sources
FSC® C105338

If you have any concerns about our products,
you can contact us on
ProductSafety@springernature.com

In case Publisher is established outside the EU,
the EU authorized representative is:
Springer Nature Customer Service Center GmbH
Europaplatz 3, 69115 Heidelberg, Germany

Printed by Libri Plureos GmbH
in Hamburg, Germany